Study Guide with Selected Solutions
to accompany
WORLD OF
CHEMISTRY
SECOND EDITION

Walt Volland

**Bellevue Community College,
Washington**

Saunders Golden Sunburst Series

Saunders College Publishing
Harcourt Brace College Publishers

Fort Worth Philadelphia San Diego New York Orlando Austin
San Antonio Toronto Montreal London Sydney Tokyo

Volland; Study Guide with Selected Solutions to accompany
<u>World of Chemistry, 2e.</u> Joesten & Wood.

ISBN 0-03-004498-7

567 021 987654321

Preface

This Study Guide is a little different from most. It is designed to be more than just a help to you in your present chemistry class. I intended that it bring some fun into the study of chemistry. I tried to make chemistry more enjoyable and less threatening and boring. I hope the activities will heighten your curiosity about the world. I included a variety of them so that you can see the many ways that all of us are connected to chemistry. The simplest, kitchen counter Bridging the Gap exercises and the more elegant Internet adventures all can lead to a wider and more understandable world. I also want you to take something of lasting value from this book. It might be the systematic methods that are part of problem solving, it might be an increased awareness of your surroundings, or it might be a continuing addiction to Internet surfing. These things can be useful for the rest of your life. Good luck in all that you do.

Why are things are way they are.

The summaries, objectives and key terms are aimed at helping you focus on the highlights contained in each section from the point of view of someone who knows, and actually enjoys chemistry. The additional readings are in all types of publications from *Mademoiselle* and *Glamour* to *The American Scientist*. When space allows, the answers to questions, include an explanation of the reasoning behind the answer. Similarly the solutions to problems are written to help you increase your analytical skills. The cross word puzzles are supposed to give you an entertaining way to build your chemistry vocabulary. The Bridging the Gap activities are aimed at showing how your everyday activities are linked to chemistry and science. The Internet activities are only a taste of the future direction of our daily lives and education. The movies like "Waterworld" remind us of the power of entertainment to shape our view of reality even though they may be fictional.

Acknowledgments

I want to thank the authors of the World of Chemistry, Mel Joesten and Jim Wood, and the Contributing Editor, Mary Castellion, for their confidence in me. A special note of appreciation goes to, Beth Rosato, Developmental Editor, for her patience, trust, and constant encouragement.

My deepest thanks go to my wife Gloria and my sons Kirk and Greg. Gloria's understanding, help at proof reading, editing, and heroic support are what made this book possible. Kirk gave invaluable advice regarding the Internet exercises and computer graphics. Greg cheered me along and helped me keep the work in perspective.

Feedback

I am interested in receiving comments and suggestions. Please contact me by mail or email.

Dr. Walt Volland
Department of Chemistry
Bellevue Community College
3000 Landerholm Circle S.E.
Bellevue, WA 98007-6484

email: wvolland@bcc.ctc.edu
fax: 206-641-2230

WVV
Bellevue, Washington
JUNE 1996

Internet Assignments

How do you get on the Internet?
Some of you already have an Internet provider; like America Online, Prodigy, CompuServe, but many people do not. If you do not have the money or desire to buy a system you still can use public Internet access facilities. Internet access is limited now but not for long. Your public library may offer access. Call and ask if library users have access to computer terminals with Internet service. Many colleges offer Internet access through the libraries, the campus computer centers, departmental terminals or even have it wired to dormitory rooms. You should ask your instructor or someone in computer services about on campus access.

A simple start on the Net
There is a tremendous amount of trivial information on the Internet. It also provides a direct quick connection to government agencies and other sources of hard factual information. The exercises in this book are designed to give you a reason to learn about the power of the Internet. The ease of contacting people and accessing information from all the corners of the world can be breathtaking. You can read an article in London one minute and then see a satellite view of weather fronts from NOAA in the next.

The Internet typically is accessed using a browser. Most browsers look alike. One of the most common browsers is Netscape© which is a very good browser and is free to students, educators and schools. The Show Location window for Netscape© is shown below. The Universal Resource Locators URLs given in the Internet assignments can be typed in the "Location: " window. The URL you see in the picture is for the United State Geological Survey.

The different command buttons can be explored if you want. For example the "Back" button lets you go back to a previously entered URL. The "Net Search" button opens up various search engines that enable you to search the net like you would your library computer search engine. The big difference is that on the Internet you are searching worldwide sources.

What is on an Internet page
The Internet pages typically have interactive or "hot" buttons. The most common interactive spots are underlined text items. Sometimes there are images that are buttons. Usually they are identified with "Click here" messages. You can't go wrong by clicking on an item to find out if it is active. ONE CAUTION, there is a time delay between the time you click and the time when the system responds and connects you. You need some patience here. Computers have gotten faster, but people are asking the computer to do more. You probably already know you can't rush the machines. If you change your mind, you can always click on the "Stop" button. This is a chance to surf the net. Go ahead; get your feet wet.

Table of Contents

1   Living in a World of Chemistry

1.1   The World of Chemistry

This section points out that all too often the word "chemical" is associated with hazardous or toxic substances. A point is made that everything whether natural or man made contains some chemical because all matter is some kind of chemical. The future of humanity and of the planet depend on the way in which chemicals are used. The definition is given for a scientific fact. The scientific method and the chemist's way of viewing the world are described. The idea of keeping a laboratory notebook or journal is introduced.

Objectives

After studying this section a student should be able to:
> give examples of scientific facts
> tell why models and theories are developed
> tell what happens when experimental results disagree with results predicted by theory
> explain how observations and facts relate to models and theories
> tell why a laboratory notebook or journal is essential in chemistry and science

Key terms

chemical	all natural products	scientific fact
scientific method	experiment	notebook
models	theories	predictions

1.2   DNA Fingerprinting, Biochemistry and Science

This section explains how the field of DNA testing, though front page news in sensational criminal cases, is dependent on biochemistry. The historical and practical links between the physical sciences and biological sciences are described. The categories of natural sciences are outlined. The overlapping roles of physical and biological science in the study of DNA are highlighted.

Objectives

After studying this section a student should be able to:
> name the two branches of natural science
> give the definition for chemistry
> name four divisions of chemistry, tell what they study
> name two disciplines of chemistry that deal with DNA testing

Key terms

natural science	physical science	biological science
chemistry	biochemistry	analytical chemistry
organic chemistry	inorganic chemistry	physical chemistry
botany	genetics	natural matter
synthesize new matter	DNA fingerprint	

     1

1.3 Air Conditioning, the Ozone Hole, and Technology

This section recounts the historical development of chlorofluorocarbons as refrigerant gases and as propellants. The benefits and unforeseen consequences of their use are described. The ozone depletion problem is explained. A brief summary is given of the time line from creation to commercial use of various inventions.

Objectives

After studying this section a student should be able to:

describe the history of chlorofluorocarbons from development to common use
give definitions for basic science, applied science, technology
describe the link between CFCs and ozone depletion
describe how the incubation period for innovations has changed since 1900

Key terms

stable	applied and basic science	CFCs and refrigerants
technology	ozone	incubation period

1.4 Teflon, Scientific Discovery, and Serendipity

This section is a case study of the discovery of Teflon. It illustrates how serendipity and careful observation can yield discoveries.

Objectives

After studying this section a student should be able to:

describe the history of the discovery and development of Teflon
explain how Plunkett's work illustrates the importance of accidental discoveries, careful observation and detailed measurements

Key terms

Teflon	CFCs	polymers

1.5 Automobile Tires, Hazardous Waste, and Risk

This section describes the case of a fire in a scrap tire dump. The risks associated with tire disposal are discussed. The concepts of risk assessment and management are explained. Estimates of increased risk of death after engaging in various activities are tabulated. Governmental actions to attempt to regulate risks are described. The Delaney clause and zero risk laws are explained.

Objectives

After studying this section a student should be able to:

give the definitions for risk assessment and risk management
give the definitions for risk-based laws, technology-based laws, and zero-risk laws
describe the Delaney clause
give the definition for a carcinogen

Key terms

spontaneous chemical change	risk assessment	risk-based laws
Delaney clause	carcinogen	balancing laws
technology-based law	risk management	EPA

1.6 What is Your Attitude Toward Chemistry?

This section discusses the reasons to study chemistry. The idea that protection of health, safety and the environment depend on a knowledge of chemistry is introduced.

Objectives
After studying this section a student should be able to:

explain why chemophobia may make matters worse

tell how a knowledge of chemistry can make decisions about health and the environment more effective

Key terms and equations
chemophobia

Additional readings

Ausubel, J. H. "Can Technology Spare the Earth?" American Scientist Mar-April 1996: p 166.

Beardsley,Tim. "Death by Analysis" Scientific American June 1995 : p 28.

Gibbs, W. Wayt , "Ounce of Prevention: Cleaner chemicals pay off, but industry is slow to invest" Scientific American Nov 1994: p 103.

Horgan, John. "Radon's Risks: Is the EPA exaggerating the dangers of this ubiquitous gas?" Scientific American Aug. 1994: p 14.

Lander, Eric S. "DNA Fingerprinting on Trial" Nature June 15, 1989: p 505.

Neufeld, Peter J. and Neville Colman. "When Science Takes the Witness Stand" Scientific American May 1990: p 46.

"Putting Sunscreens to the Test" Consumer Reports May 1995: p 334.

"Secondhand Smoke, Is It a Hazard?" Consumer Reports Jan. 1995: p 27.

Zorpette, Glenn. "Bracing for the Next Big One: Engineers grapple with retrofitting Japanese and U.S. buildings" Scientific American April 1995: p14.

Speaking of Chemistry

Name _____

Living in a World of Chemistry

Across

2 An unreasonable fear of chemicals.
4 The father of synthetic chemistry. An English scientist who discovered a way to make synthetic dyes when trying to make quinine.
6 Deoxyribonucleic acid
7 Artificial or man made duplicate of a naturally occurring substance.
9 Household ammonia mixed with chlorine _____ can produce a toxic gas.
12 Gases like CFCs used in refrigerator and air conditioner compressors.
14 Gas in the upper atmosphere that is being depleted by CFCs.
15 An observation that does not change like the boiling point of water at 100°C.
16 A medicine used to treat malaria.

Down

1 Polymer accidentally discovered by Plunkett in 1938.
2 Chlorofluorocarbons
3 Interpretations summarizing data.
5 _____ period, the time interval between the conception of an idea and its realization.
8 Experimental observations
10 Branch of chemistry that deals with carbon compounds.
11 Condition where no risks exist.
12 A chance or gamble
13 Clause in the Food, Drug and Cosmetic Act that bans use of any carcinogenic additives in processed food.

Bridging the Gap I
Concept Report Sheet

Name _____

Risks: DNA Fingerprinting and Secondhand Smoke

This is a two part exercise. The first part addresses the issue of the admissibility of DNA fingerprinting as evidence in criminal cases and the second deals with secondhand smoke.

1.   Admissibility of DNA Fingerprinting evidence

Judicial decisions to admit laboratory test results as evidence in criminal cases are fraught with problems and risks for society. The traditional attitude in the criminal justice system in the United States accepts legal precedence as a basis for admissibility of evidence. Perhaps the most notorious example of the problem this poses is the so-called paraffin test which was used by crime laboratories throughout the U.S. to detect nitrite and nitrate residues, presumably from gunpowder, on suspects' hands to show that they had recently fired a gun. The test was first admitted as scientific evidence in a 1936 trial in Pennsylvania. Other states then adopted that decision without scrutinizing the supporting research. For 25 years numerous defendants were convicted with the help of this test. It was not until the mid-1960's that the test was subjected to careful scientific study; the study revealed a damning flaw in the test. It gave an unacceptably high number of false positives. Substances other than gunpowder that gave a positive reading included urine, tobacco, tobacco ash, fertilizer and colored fingernail polish. In this instance the legal process failed, allowing people to be convicted on scientific data that later proved to be worthless.

Pretend you are a judge. Develop two questions you would want answered by experts before you would decide whether or not DNA fingerprinting should be admissible as evidence in criminal cases. Include a reason for wanting each of these questions answered. Write these questions and reasons on this sheet.

2. Secondhand Smoke

Decisions in daily life are made based on many factors. One such decision deals with the question of smoking. Personally each person can make their own choice, but the issue of secondhand smoke has created pressure for legislation to bar smoking in public places. Pretend you are a member of the state legislature. Your task is to develop two questions you would want answered by experts regarding the risks of secondhand smoke before you would decide on legislation. Include a reason for wanting each of these questions answered. Write these questions and reasons on this sheet.

Bridging the Gap   II

Internet access to the
Center for Disease Control, CDC and the
Federal Drug Administration, FDA

There is a tremendous amount of trivial information on the Internet. It also provides a direct quick connection to government agencies and other sources of hard factual information. This is an exercise in using the Internet to learn about the power of the Internet. Simultaneously you will have an opportunity to learn what health agencies like the FDA and the CDC say about health risks and smoking.

The Internet pages typically have interactive or "hot" buttons. The most common interactive spots are underlined text items. Sometimes there are images that are buttons. Usually they are identified with "Click here" messages. You can't go wrong by clicking on an item to find out if it is active. ONE CAUTION, there is a time delay between the time you click and the time when the system responds and connects you. You need some patience here. Computers have gotten faster, but people are asking the computer to do more. You probably already know you can't rush the machines. This is a chance to surf the net. Go ahead; get your feet wet.

The CDC home page and the FDA

Public agencies like the Center for Disease Control, CDC, and the Food and Drug Administration, FDA, can be accessed through the Internet. The CDC home page can be reached by using the following universal resource locator (URL) web site address.

> http://www.cdc.gov/

The Center for Disease Control has an information page that includes data about risks and tobacco use. On the CDC information page go to the line reading "Diseases, Health Risks, Prevention Guidelines and Strategies", and click on "Health Risks"; then click on "Behavioral Risk Factors" and go to "Tobacco Use". Alternatively, the CDC Tobacco Use Information Page can be reached directly by using the following URL instead of going through the CDC information page.

> http://www.cdc.gov/nccdphp/osh/tobacco.htm

Your assignment is to open this page. It will have a heading like the one shown below.

·CDC Tobacco Use Information Page

You are to find out what the CDC says about the number of deaths annually in the United States that are said to be caused by smoking. Record your information on the Concept Report Sheet.

In addition, you are to open the active line that gives the Food and Drug Administration's proposed regulations on tobacco by clicking on the underlined FDA entry. If you have trouble, you can use the following URL for the FDA site. You are supposed to record your opinion of the FDA proposals on tobacco regulation and give a reason for your opinion.

> http://www.fda.gov/opacom/campaigns/tobacco.html

Bridging the Gap   II   Name _____

Internet access to the
Center for Disease Control, CDC and the
Federal Drug Administration, FDA

1.   What is the Center for Disease Control's estimate of the number of deaths in the United States that are annually caused by tobacco. Does this seem like a large number to you? Why?

2.   What is your opinion of the regulations proposed by the FDA ? Would you endorse these proposals? Explain.

3.   Give the universal resource locator (URL) for another interesting site you found while doing this exercise.

   http://_____

   What makes this site noteworthy?

10

Solution to Speaking of Chemistry crossword puzzle

Living in a World of Chemistry

			¹T									
²C	H	E	M	O	P	H	O	B	I	A		
F		F					³C					
C		L	⁴P	E	R	K	I	N				
		O			⁵N		⁶D	N	A			
⁷S	Y	N	T	H	E	T	I	C			⁸D	
					U		⁹B	L	E	A	C	H
	¹⁰O		¹¹Z	B		U			T			
¹²R	E	F	R	I	G	E	R	A	N	T	S	A
I	G		R		T		S		A		¹³D	
S	A		O		I		¹⁴O	Z	O	N	E	
K	N		R		O		N				L	
	I		I		N		S				A	
¹⁵F	A	C	T								N	
			S		¹⁶Q	U	I	N	I	N	E	
			K								Y	

2   The Chemical View of Matter

2.1   Elements -- the Most Simple Kind of Matter

Main topics in this section are the properties of elements, the states of matter, how the properties of elements can be used to distinguish one from another, and the chemical model of matter.

Objectives

After studying this section a student should be able to:

describe the characteristics that distinguish elements from other substances
describe the type of particles that make up an element
identify the states of matter
write the definitions for a pure substance, element, and atom

Key terms

atoms	states of matter	solids	liquids
elements	gases	plasmas	pure substances

2.2   Chemical Compounds -- Atoms in Combination

Main topics in this section are how to read formulas, the characteristics of compounds, and the definition of chemical compounds.

Objectives

After studying this section a student should be able to:

use the chemical formula to tell what elements make up a compound
identify the two types of pure substances
describe the characteristics that distinguish compounds from other substances
describe the type of particles that make up a compound

Key terms

chemical compound	molecule	subscript
decomposition	element	characteristic property

2.3   Mixtures and Pure Substances

Main topics in this section are the classes of mixtures, descriptions of homogeneous and heterogeneous mixtures, and description of methods for the separation of mixtures.

Objectives

After studying this section a student should be able to:

name the types of mixtures
tell how homogeneous mixtures and heterogeneous mixtures differ
describe what happens to a mixture when it is purified
explain how purification of a mixture leads to a change in properties
describe ways to purify common mixtures such as

Key terms

mixtures	heterogeneous mixtures	homogeneous mixtures
solutions	separation of mixtures	

13

2.4   Changes in Matter: Is It Physical or Chemical

This section discusses the definition and description of physical properties, definition of density, physical change, chemical properties, chemical reaction, forms of energy, potential energy and kinetic energy.

Objectives

After studying this section a student should be able to:

identify physical properties
give the definition for density
identify chemical properties
distinguish between physical and chemical properties
give the definitions for energy, potential energy and kinetic energy

Key terms

physical properties	physical change	physical separation
density	molecule	chemical property
chemical reaction	reactants	products
energy	potential energy	kinetic energy

2.5   Classification of Matter

Main ideas in this section are the descriptions of homogeneous mixtures, heterogeneous mixtures, pure substances, major reasons for studying pure substances, and the connection between the structure of matter and chemical and physical properties.

Objectives

After studying this section a student should be able to:

describe the structure and properties of a homogeneous mixture
describe the structure and properties of a heterogeneous mixture
give reasons for studying pure substances

Key terms

heterogeneous	homogeneous	structure of matter

2.6   The Chemical Elements

This section includes a discussion of the classes of elements and their symbolic representation.

Objectives

After studying this section a student should be able to:

describe what is meant by naturally occurring elements
tell what an artificial element is and
describe what is meant by the term radioactive
describe the properties of metals and nonmetals
identify eight of the most common nonmetals
tell what is meant by the term diatomic molecule in regard to the elements

Key terms

naturally occurring elements	artificial elements	radioactive
metals	nonmetals	common nonmetals
diatomic molecules	symbol	

2.7 Using Chemical Symbols

Major topics developed in this section are types of chemical formulae, characteristics of organic and inorganic compounds, information available from chemical equations, and how to balance chemical equations.

Objectives

After studying this section a student should be able to:

read a formula for a compound and tell the identities of the elements
tell the relative numbers of atoms of each element from a chemical formula
tell how structural formulas, molecular formulas, and condensed formulas differ
explain what information is given by a structural formula
state the definitions for organic compounds and inorganic compounds
describe the information provided by a chemical equation
identify reactants, products and a balanced chemical equation
balance an unbalanced chemical equation
distinguish between coefficients in a chemical equation and formula subscripts

Key terms

chemical formulas	subscripts	structural formulas
molecular formulas	condensed formulas	organic compounds
inorganic compounds	chemical equations	balanced equations
coefficient	aqueous solution	

2.8 The Quantitative Side of Science

This section describes the difference between qualitative and quantitative, significant digits, the metric system, fundamental units, derived units, prefixes in the metric system, techniques used in converting one metric unit to another, examples showing the use of conversion factors and the difference between mass and weight.

Objectives

After studying this section a student should be able to:

tell how qualitative and quantitative information differ and give an example of each
identify the seven fundamental physical quantities measured in SI units
give the names and symbols for the fundamental SI units
tell how a fundamental unit differs from a derived unit and give an example of each
name common metric prefixes, give their abbreviation and meaning i.e. nano-, n, 10^{-9}
give symbols for common units i.e. Meter, Liter, Gram, Degrees Celsius, Calorie.
tell how weight and mass differ
use the unit conversion method to convert one metric unit into another
use the unit conversion method to convert a metric unit to an English or American unit
give the definition of density and calculate density from mass and volume data

Key terms

qualitative	quantitative	metric system
fundamental units	derived units	physical quantity
length	amount of substance	time
electric current	scientific notation	thermodynamic temperature
meter = m	weight and mass	prefixes
mega = M = 10^6	kilo = k = 10^3	deci = d = 10^{-1}
centi = c = 10^{-2}	milli = m = 10^{-3}	micro = μ = 10^{-6}

$$\text{nano} = \text{n} = 10^{-6} \qquad \text{pico} = \text{p} = 10^{-12} \qquad \text{degrees Celsius} = {}^{\circ}C$$
$$\text{liter} = \text{L} \qquad \text{gram} = \text{g} \qquad \text{Calorie} = \text{cal}$$

Additional readings

Freedman, David H. "The Biggest Chill: Absolute Zero." Discover Feb. 1993: 62.

Hoffman, Roald. "How Should Chemists Think?" Scientific American Feb. 1993: 66.

Itano, Wayne M. and Norman F. Ramsey. "Accurate Measurement of Time." Scientific American July 1993: 56.

Leutwyler, Kristin. "Chiller Thriller: Workers Achieve Temperatures Below Absolute Zero." Scientific American Jan. 1994: 24.

Lipkin, Richard. "Physicists Spot Element 111." Science News 7 Jan. 1995: 5.

Nauenberg, Michael et al. "The Classical Limit of an Atom." Scientific American June. 1994: 44.

Nelson, Robert A. "Guide for Metric Practice." Physics Today Aug. 1995: 15.

Raloff, Janet.. "Microwaves Accelerate Chemical Extractions." Science News 21 Aug. 1993: 118.

Seligman, Daniel. "Metric mania." Fortune 19 Oct. 1992: 132..

Zebrowski, George. "Time is nothing but a clock." Omni Oct. 1994: 80.

Answers to Odd Numbered Questions for Review and Thought

1. Materials typically used in an average day that were not chemically changed from their natural states are gold, oxygen in the air, water methane in natural gas, beeswax, wood.

3. No. Two pure substances that have the same set of physical and chemical properties would be the same. Two different substances will have different formulas or structures. This typically produces different physical and chemical properties. Some properties may be the same but it is highly unlikely that all would be.

5. When a BIC lighter is lit the spark provides the energy needed to raise the temperature of the air and BIC fuel mixture above the ignition temperature. The mixture ignites and bursts into flame. The mixture of oxygen, $O_{2(gas)}$ and hydrocarbon fuel are chemically changed to heat and new substances water vapor, $H_2O_{(gas)}$, and carbon dioxide, $CO_{2(g)}$.

7. a. Density is a physical property because it depends only on the physical properties of mass and volume. d = mass/ volume
   b. Melting temperature is a physical property. It is linked to the amount of energy needed to separate the particles in a solid and allow them to move freely in the liquid.
   c. Decomposition of a substance into two elements upon heating is a chemical property because the substance changes from its original formula to two elements. There is a change in chemical composition.
   d. Electrical conductivity is a physical property. The chemical composition of a piece of copper wire does not change when it conducts electricity.

e. The failure of a substance to react with sulfur is a chemical property. The chemical composition is unchanged when no reaction occurs.

f. The ignition temperature of a piece of paper is a chemical property. It indicates the ease with which the paper reacts with air resulting in a change in composition.

9. a. Mercury is an element. Division of a sample of mercury leads to separate Hg atoms.
   b. Milk is a mixture of water, minerals, proteins, fats.
   c. Pure water is a compound, contains only one kind of molecule (H_2O)
   d. Wood is a mixture of cellulose, water. Wood changes weight when dried and loses H_2O.
   e. Ink is mixture of dye and solvent. There is more than one type of molecule in mixture.
   f. Iced tea is a mixture of water, caffeine, tea extract. More than one type of molecule
   g. Pure ice compound, solid pure water containing only one kind of molecule, (H_2O).
   h. Carbon contains one kind of atom. It is an element, it contains only C atoms.
   i. Antimony contains one kind of atom. It is an element, it contains only Sb atoms.

11. Properties of iron do not change because all particles in iron are atoms of iron. Steel is an alloy or mixture of iron and other atoms. The type of steel depends on what is added.

13. The water can be evaporated or distilled off, trapped and condensed. The solids will be left behind as a residue.

15. Atrazine, $C_8H_{14}N_5Cl$, contains carbon, C, hydrogen, H, nitrogen ,N, chlorine, Cl.

17. a. Calcium carbonate, $CaCO_3$,   Bon Ami kitchen and bath cleanser
    b. Phosphoric acid, H_3PO_4,   Coca Cola
    c. Water, H_2O,   Gatorade
    d. Fructose, $C_6H_{12}O_6$,   Coca Cola
    e. Sodium chloride, NaCl,   Skippy Peanut Butter
    f. Potassium sorbate, $KC_6H_7O_2$,   Kraft Grated Parmesan Cheese
    g. Potassium iodide, KI,   Morton's Iodized Salt
    h. Glycerol (glycerin), $C_3H_8O_3$,   Oil of Olay
    i. Aluminum is typically in a
       compound in most consumer products.
       Aluminum hydroxide, $Al(OH)_3$,   Mylanta
    j. Butylated hydroxytoluene, (BHT), $C_{15}H_{24}O$,   Kellogg's Frosted Mini-Wheats

19. N_2 + 3 H_2 $\longrightarrow$ 2 NH_3

21. electrical, heat, light, mechanical

23. a. On the left side of the arrow, "2 Na" means 2 Na atoms; one Cl_2 molecule contains 2 Cl atoms. On the right side 2 NaCl units contain 2 Na atoms and 2 Cl atoms.

b. On the left one N_2 molecule contains 2 N atoms and 3 Cl_2 molecules contain 6 Cl atoms. On the right 2 NCl_3 molecules contain a total of 2 N atoms and 6 Cl atoms.

c. On the left there are 1 C atom, 2 H atoms and 2 + 1 = 3 O atoms. On the right there are 1 C atom, 2 H atoms and 3 O atoms.

d. On the left there are 4 H atoms and 4 O atoms in 2 molecules of H_2O_2. On the right there are 4 H atoms in the 2 molecules of water; there are also 2 O atoms in the 2 water molecules and 2 more O atoms in the O_2 molecule for a total of 4 O atoms.

25. a. No.   b.   Yes.

27. Most substances in the kitchen are mixtures of compounds. A few pure ones are baking soda, sugar, flour, distilled water, aluminum foil

29. Mass measured in grams. Length is measured in meters. Volume is measured in liters.

31. Yes. A liter is roughly equal to a quart. The actual relationship is 1 quart = 0.9463 liter.

Solutions for Selected Problems

1. Remember the definition for a dozen can be used as a conversion factor. All equalities can be used in this way. Conversion factors depend on the cancellation of units. Words in the defined units cancel just like numbers cancel. $\dfrac{21 \times 2}{2} = 21$; 1 dozen = 12 marbles

   ? marbles = $\dfrac{8 \text{ dozen marbles}}{1} \times \dfrac{12 \text{ marbles}}{1 \text{ dozen marbles}}$ = 96 marbles

2. The question asks for the number of 200 ml glasses that can be filled using 10 liters of liquid. Remember 1 liter = 1000 ml this gives the conversion factor $\dfrac{1000 \text{ ml}}{1 \text{ liter}}$. This can be used to convert any volume in units of liters to ml.

   $\dfrac{10 \text{ liter}}{1} \times \dfrac{1000 \text{ ml}}{1 \text{ liter}} \times \dfrac{1 \text{ glass}}{200 \text{ ml}}$ = 50 glasses

3. Remember the prefix definitions of mega and kilo. mega = M = 1,000,000 = 1×10^6
kilo = k = 1000 = 1×10^3

$$? \text{ kilobytes} = \frac{200 \text{ Mbytes}}{1} \times \frac{1,000,000 \text{ bytes}}{1 \text{ Mbyte}} \times \frac{1 \text{ kilobytes}}{1000 \text{ byte}} = 200,000 \text{ kilobytes,}$$

$$? \text{ bytes} = \frac{200 \text{ Mbytes}}{1} \times \frac{1,000,000 \text{ bytes}}{1 \text{ Mbyte}} = 200,000,000 \text{ bytes}$$

4. A normal serving of peanut butter has a mass of 36 grams. The reduced fat type has 30% less fat than the normal serving and contains 12 g of fat. This means the 12 g of fat is 70% of the normal amount of fat in a serving of regular peanut butter.
Let y = g of fat in regular peanut butter.

$$12 \text{ g fat} = \frac{70}{100}y \; ; \; 12 \text{ g fat} = 0.70 \, y \; ; \quad y = \frac{12 \text{ g of fat}}{0.70} = 17 \text{ g of fat.}$$

5. The answer is supposed to be in units of acres per cow. The relation between acres and cows is 1 acre/10 cows. This means there would be 5 acres/50 cows or 6 acres/60 cows. Multiply top and bottom by 5.5.

$$\frac{? \text{ acres}}{55 \text{ cows}} = \frac{1 \text{ acre}}{10 \text{ cows}} \times \frac{5.5}{5.5} = \frac{5.5 \text{ acres}}{55 \text{ cows}} \text{ or } 5.5 \text{ acres/55 cows}$$

6. The mass of an Advil tablet = 200 mg ibuprofen. ; Converting the mg to grams uses the link between grams and milligrams. 1000 mg = 1 g

$$\frac{200 \text{ mg ibuprofen}}{1 \text{ tablet}} \times \frac{1 \text{ gram}}{1000 \text{ mg}} = \frac{0.200 \text{ gram ibuprofen}}{1 \text{ tablet}}$$

The conversion of micrograms to grams is needed for the second part. 1,000,000 μg = 1 g

$$\frac{200 \text{ mg ibuprofen}}{1 \text{ tablet}} \times \frac{1 \text{ gram}}{1000 \text{ mg}} \times \frac{1,000,000 \text{ μg}}{1 \text{ gram}} = \frac{200,000 \text{ μg ibuprofen}}{1 \text{ tablet}}$$

7. Volume is a derived unit. It can be calculated from the dimensions of a regular box by using the relationship V = length x width x height = l x w x h
Volume = 20 cm x 10 cm x 5 cm = 1000 cm^3

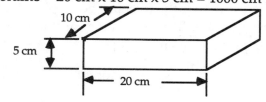

8. The ratio of mass to volume is density. Density is an intensive property that is characteristic of the substance and can be used to identify a compound. Density is a derived unit that depends on the fundamental unit mass and derived unit of volume. The symbol for density is D and the formula for calculating density is

$$D = \frac{\text{mass in grams}}{\text{volume in cm}^3}.$$ Substitute 3000 grams for mass and 1000 cm^3 for volume and

calculate answer. $$D = \frac{3000 \text{ gram}}{1000 \text{ cm}^3} = 3. \text{ g/cm}^3$$

9. You need to use the conversion factors 1 m = 100 cm and 1 kg = 1000 g in order to convert the density units from g/cm^3 to kg / m^3.

$$\#kg/m^3 = \frac{3\text{ g}}{cm^3} \times \frac{1\text{ kg}}{1000\text{ g}} \times \frac{100\text{ cm}}{1\text{ meter}} \times \frac{100\text{ cm}}{1\text{ meter}} \times \frac{100\text{ cm}}{1\text{ meter}} = \frac{3\text{ g}}{cm^3} \times \frac{1\text{ kg}}{1000\text{ g}} \times \left(\frac{100\text{ cm}}{1\text{ meter}}\right)^3 =$$

$$\frac{3\text{ g}}{cm^3} \times \frac{1\text{ kg}}{1000\text{ g}} \times \frac{1{,}000{,}000\text{ cm}^3}{1\text{ meter}^3} = 3000\text{ kg/m}^3$$

13. a. Use the definition 100 cm = 1 m; 0.04 m = 4 cm $\times \dfrac{1\text{ meter}}{100\text{ cm}}$

   b. Use the conversion factor 1000 mg = 1 g; 43 mg = 0.043 g $\times \dfrac{1000\text{ mg}}{1\text{ g}}$

   c. Use the conversion factor 1000 mm = 1 m; 15500 mm = 15.5 m $\times \dfrac{1000\text{ mm}}{1\text{ meter}}$

   d. Use the conversion factor 1000 ml = 1 L ; 0.328 L = 328 ml $\times \dfrac{1\text{ L}}{1000\text{ ml}}$

   e. Use the conversion factor 1000 g = 1 kg ; 980 g = 0.980 kg $\times \dfrac{1000\text{ g}}{1\text{ kg}}$

18. Density is defined by the relation; D = $\dfrac{\text{mass in grams}}{\text{volume in cm}^3}$. and solving for mass the result is mass = D x volume. Here the volume is 50.0 cm^3 and the density is 19.3 g/cm^3.

   965 g = $\dfrac{19.3\text{ g}}{cm^3}$ x 50.0 cm^3.

20. Density is defined by the relation; D = $\dfrac{\text{mass in grams}}{\text{volume in cm}^3}$. and solving for volume the result is volume = $\dfrac{\text{mass in grams}}{\text{density}}$. The volume is mass 500 ml

22. The comparison of temperatures can be made because the freezing point of water in the Fahrenheit scale is 32°F and 0°C in the Celsius scale. This makes it clear that 0 °F is colder than 0°C because 0 °F is below the freezing point for water.

Speaking of Chemistry

Name _____

Chemical View of Matter

Across

2 Metric prefix meaning million.
6 Temperature scale with boiling point for water at 100
8 Indicated by the prefix mega-.
12 Compounds and elements are the two classes of _____ substances.
13 Smallest particles of an element
16 Mixture that is not uniform.
17 Multiple indicated by prefix micro-.
20 Metric unit equal to 100 cm.
21 Physical property measured in meters.
22 Formulas stay the same in a _____ change.

Down

1 Prefix meaning 1/10.
2 Property measured by grams and kg.
3 Metric unit equal to 1/1000 kg.
4 ____ units like volume are developed from fundamental units.
5 Metric prefix meaning 1/100.
7 Metric unit meaning 1000.
9 Class of compounds that are not organic.
10 The most compressible state of matter.
11 Derived metric unit for volume.
13 Solution with water as the solvent.
14 Particle consisting of 2 or more atoms.
15 Elements like lead and aluminum.
17 Metric prefix meaning 1/1000.
18 Property measured in units of seconds.
19 _____ is the ability to cause change or do work.

Bridging the Gap   I
Separation of Mixtures, Paper Chromatography

Mixtures are either homogeneous or heterogeneous. Homogeneous mixtures are uniform in composition. Heterogeneous mixtures are not. Salt water is a solution of water and NaCl. and is homogeneous if thoroughly mixed. Both types of mixtures can be separated into by physical means. Food colorings are typically a homogeneous mixture of a solvent a single dye or a combination of selected dyes that produce the desired color. You will use paper chromatography to test a food coloring to see if it is a pure substance of a single dye or mixture of dyes and solvent. Other substances that are not colored can be detected using ultraviolet or black light. These substances appear to glow in the dark. Your exercise is simpler but uses very essential principles.

Equipment and materials
Clear colorless glass or plastic tumbler or jelly jar, paper coffee filter, toothpicks, scissors, pencil, adhesive tape, tap water and a package of Schilling® or other brand of food coloring.

1.   Cut a half inch wide (1.25 cm) strip of coffee filter paper about four inches long(10 cm).

2.   Make a start line with pencil mark at half an inch from one end that will be the bottom.

Top   Bottom

Pencil marked start line

3.   Cut off one end of a toothpick. Dip the fresh cut end into the food coloring.

4.   Use the toothpick to place a dot of blue food coloring on the pencil mark start line and allow it to dry.

5.   Attach a piece of tape to the top end of the strip of paper. Tape the paper to the pencil and lower the paper into the tumbler. Check how far the paper projects into the container. Remove the pencil and paper.

6   Add water to container so the water level will touch the paper 1/2 inch below the start line.

7.   Lower the paper into the tumbler so the water touches the end of the paper at least 1/2 inch below the spot.

8.   Let the water wick up the paper. Note where the water wets the paper, this is the solvent front. The water will climb the first inch quickly. The dye will trail behind the water.

9.   When the front edge of the water reaches half way up the paper remove the paper from the glass. Allow the paper to dry. Note if more than one color appears on the paper.

10.   Repeat steps 1 through 8 but using the yellow dye and then the green dye. Record your observations on the Concept Report Sheet. Which of the dyes are single colors and which are mixtures. Explain your reasoning.

           23

Bridging the Gap   I   Name _____
Concept Report Sheet
Separation of Mixtures, Paper Chromatography

Data

Food coloring	Colors observed
Blue	
Yellow	
Red	
Green	

Food coloring	Mixture Yes/ No	How many components are in the dye.
Yellow		
Blue		
Red		
Green		

Concepts and analysis

Which of the food coloring dyes is a mixture? Explain your reasoning.

Do you think that other substances like vegetable dyes or inks could be tested using this method? Explain.

Explain briefly what would have to be done if the dyes would not dissolve in water? .

Name an everyday activity that involves the separation of a mixture. For example tea is brewed using hot water to extract desirable substances from the tea leaves.

Bridging the Gap   II   Name _____

"To be or not to be" The English or the Metric system

The English system of measurement used today in the United States originated in the decrees of English monarchs. The French Revolution produced the overthrow of the French monarchy and in 1799 it also created of the set of weights and measures we call the metric system. The metric system was legalized for use in the United States in 1866 along with the traditional English system. Today the only countries in the world that do not use the metric system are the United States of America, Liberia and Myanmar.

This exercise has three parts. One part is to write a brief argument for adopting the metric system and replacing the English system. Another part of your assignment is to write an argument for continuing the current pattern. Lastly you are to write two questions you would want answered before you would make a decision about this issue.

Concept Report Sheet

1.   Argument for adopting the Metric system and replacing the English system

2.   Argument for retaining both the English system and the Metric system

3.   Your questions regarding the issues.

Solution to Speaking of Chemistry crossword puzzle
Chemical View of Matter

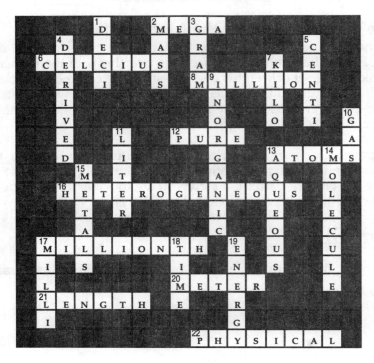

3   Atoms

3.1   The Greek Influence on Atomic Theory
The main topic in this section is the historical development of the theory of matter. Both the continuous theory proposed by Plato (427-347 BC.) and Aristotle (384-322 BC.) and the atomic theory proposed by Democritus and Leucippus, and later refined by John Dalton in 1803, are presented.

Objectives
After studying this section a student should be able to:
> describe the continuous view of matter proposed by Plato and Aristotle
> describe the view of matter proposed by Democritus and Leucippus
> give the time period when Democritus proposed his theory of matter
> give the time period when John Dalton proposed his theory of matter

Key terms
atomic theory	Democritus	Leucippus
continuous structure	Plato	Aristotle
Dalton		

3.2   John Dalton's Atomic Theory
Main topics described in this section are the provisions of Dalton's atomic theory, the law of conservation of matter, the law of definite composition, the law of multiple proportions, an example of the operation of the law of definite composition and the personal histories of John Dalton and the father of modern chemistry, Antoine-Laurent Lavoisier

Objectives
After studying this section a student should be able to:
> describe and state the provisions of Dalton's atomic theory
> state the three scientific laws that Dalton's atomic theory attempted to explain
> state and describe the law of conservation of matter
> state and describe the law of definite composition
> state the law of multiple proportions
> show how the law of multiple proportions applies to a specific pair of elements

Key terms
quantitative	Dalton	law of conservation of matter
atoms	Lavoisier	law of definite composition
law of multiple proportions		

3.3   Structure of the Atom
This section deals with the origins of radioactivity, descriptions of historic experiments that revealed the properties of subatomic particles, the roles of scientists like Franklin, Curie, Thomson, Millikan, and Rutherford. It describes the properties of electrons, protons, neutrons, and Rutherford's experiment that led to the nuclear model of the atom.

Objectives
After studying this section a student should be able to:

tell how a radioactive element differs from a nonradioactive one

describe the phenomenon known as radioactivity

describe Berquerel's and Marie Curie's role in the study of radioactivity

describe how like charged particles interact and give an example of this behavior

describe how unlike charged particles interact and give an example of this behavior

sketch and label a diagram of a cathode ray tube

tell what J.J. Thomson's role was in studying the electron

give a modern day application of the cathode ray tube

name the three subatomic particles and tell who discovered each

describe relative masses and charges for the proton, electron and neutron

describe the physical properties of alpha particles, beta particles and gamma rays

sketch and label the equipment set up for Rutherford gold foil experiment

describe the observations and interpretation of Rutherford's gold foil experiment

Key terms

natural radioactivity	Marie Curie	subatomic particles
atomic structure	electrical charge	like charges
cathode rays	cathode ray tubes	cathode and anode
electrically neutral	electrons	charge to mass ratio
nucleus	ion	protons
beta particle, β	neutrons	alpha particle, α
gamma, γ, rays	Rutherford's model	Rutherford's experiment

3.4   Modern View of the Atom

This section discusses the locations of the subatomic particles in the atom, the methods to measure positions of atoms like x-ray diffraction, the scanning tunneling microscope, the atomic force microscope. The relative size of atoms, and the relation of atomic number and mass number to atomic structure are explained.

Isotopes, their relative abundances, atomic weights, the atomic mass unit, average atomic weight are described. An introduction to reading periodic table entries is given.

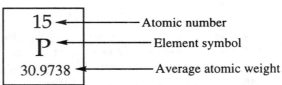

Objectives
After studying this section a student should be able to:

tell what x-ray diffraction, STM and ATM are used for in the study of atoms

tell how the number of electrons and the number of protons are related in a neutral atom

describe the locations for electrons, protons and neutrons in an atom

give the definition for atomic number, mass number and isotopes

explain how to determine the neutron number from mass number and atomic number

identify a pair of isotopes from information and notation like $^{21}_{10}Ne$, $^{22}_{10}Ne$, $^{23}_{11}Na$, $^{19}_{9}F$.

tell how the atomic mass unit is defined

describe what abundance means when describing isotopes and average atomic weight

calculate the weighted average atomic mass from percent abundances and isotope masses

read the periodic table to identify an element with a specific atomic number

Key terms

X-ray diffraction	scanning tunneling microscope	STM
atomic number	atomic force microscope	ATM
mass number	isotopes	mass spectrum
relative abundance	atomic mass unit, amu	atomic weight

3.5   Where Are the Electrons in Atoms?

This section includes descriptions of emission line spectra, continuous spectra, the interaction between a light beam and a prism, the relationship between wavelength, color, frequency and energy.   The Bohr model for the hydrogen atom and the idea of energy levels are explained. Quantum theory, ground states, and excited states are introduced. Examples of electron arrangements for atoms, valence electrons, atomic orbital shapes, and electron probability contours for "s" and "p" orbitals are explained and illustrated with examples.

Objectives

After studying this section a student should be able to:

describe the continuous spectrum and how it is produced

tell how an emission line spectrum is produced

tell how an emission line spectrum differs from a continuous spectrum

describe how wavelength, energy and frequency are related to the colors of visible light

describe the role of flame tests in the identification of an element

describe what is meant by an energy level and state the restrictions on the values for "n"

explain how the ground state for an atom relates to an excited state

write out the ground state Bohr electron arrangement for neutral atoms with atomic number 1-20

predict the limit on the electron population for an energy level using the $2n^2$ rule

identify valence electrons in an atom and describe their importance

describe the difference between Bohr orbits and electron probability

sketch the orbital shapes for "s" and "p" atomic orbitals

match an electron configuration like $1s^2 2s^2 2p^3$ to an element

write the electron configuration for elements 1-20

Key terms

emission line spectrum	continuous spectrum	white light
characteristic lines	visible light	electromagnetic spectrum
X-rays	gamma rays	ultraviolet radiation
infrared radiation	microwave	quantum
frequency, ν	wavelength, λ	$\nu \lambda = c$
ground state	excited state	principal quantum number
emitted light	flame tests	energy levels or shells
valence electrons	electronic arrangement	electron configuration
$2n^2$	electron cloud	maximum number of electrons
probability	wave properties	deBroglie's hypothesis
s, p, d, f orbitals	sublevels	

Additional readings

Binnig, Gerd , & Heinrich Rohrer. "The Scanning Tunneling Microscope" <u>Scientific American</u> Aug. 1985: 50.

Boslough, John. "Worlds within the Atom" <u>National Geographic</u> May 1985: 634.

Conkling, John A. "Pyrotechnics" <u>Scientific American</u> July 1990: 96.

Connes, Pierre. "How Light Is Analyzed" <u>Scientific American</u> Sept. 1968 : 2 .

da C. Andrade, E.N. "The Birth of the Nuclear Atom" <u>Scientific American</u> Nov. 1956: 93.

Crommie, M. F. et al. "Confinement of Electrons to Quantum Corrals on a Metal Surface" <u>Science</u> 8 Oct. 1993: 218.

Naeye, Robert. "An Island of Stability" <u>Discover</u> Aug. 1994: 22.

Nassau, Kurt "The Causes Of Color" <u>Scientific American</u> Oct. 1980: 124 .

Nauenberg, Michael et al. "The Classical Limit of an Atom" <u>Scientific American</u> June. 1994: 44.

"Goldfinger" (gold particles used to pick up fingerprints) <u>Discover</u> Mar. 1992: 16.

Answers to Odd Numbered Questions for Review and Thought

1. Matter is neither created nor destroyed. Examples: flash bulb before and after use, hard boiling an egg, yarn knitted into clothing, melting ice cubes in a glass. There is no change in mass during the chemical changes (first two) or the physical changes.

3. Matter could be divided into smaller and smaller particles with each successive piece duplicating the properties of the previous size particle. The theory assumes that matter is continuous.

5. a. Dalton's atoms were indestructible. This means the atoms existing before a reaction continue to exist afterwards. The atoms react to form compounds by forming newlinks and combinations.
   b. Dalton's model of compounds had specific combinations of atoms. He proposed that each compound would have a fixed number of atoms of each element in a formula unit. These units would not differ. This requires a constant composition for a compound.
   c. Atoms combine in whole number amounts to form molecules of compound; see #7. Dalton would predict that the different combinations of carbon and oxygen in CO and CO_2 are characteristic for each compound. The ratios of masses of oxygen to a fixed mass of carbon in these compounds would be related in the ratio of small whole numbers. There would be two parts oxygen per 1 part carbon in CO_2 to one part oxygen per 1 part carbon in CO.

7. The Law of Multiple Proportions says that in two different compounds consisting of the same two elements, the masses of one element combined with a fixed mass of the second element are related by the ratio of small whole numbers. In SO_2 there are 32g S with 32g O and in SO_3 there are 32g S with 48g O; with S fixed at 32g , the ratio of the O amounts is 48g/32g or 3x16/2x16 or 3/2 in small whole numbers.

9. Cathode rays are streams of high energy electrons produced in an evacuated tube.

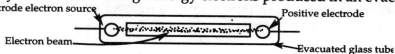

11. Ground state for an atom is the condition where all electrons are in the lowest possible energy levels. An excited state is a condition where one or more electrons are in energy levels greater than the lowest ones

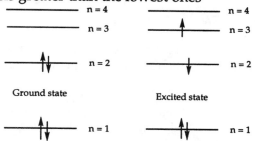

13. a. The number of protons in the nucleus of an atom.
    b. The total count of protons <u>plus</u> neutrons in the nucleus of an atom.
    c. The mass of an average atom of an element compared to an atom of ^{12}C which is assigned a mass of exactly 12 atomic mass units.
    d. Atoms with the same number of protons but with different numbers of neutrons.
    e. On Earth, the % that this isotope is of all the atoms of that element
    f. 1 amu is one-twelfth of the mass of the carbon-12 atom. see also 13.c.

15. a. No, water evaporating from a puddle merely escapes into the atmosphere to increase water content of the atmosphere. The molecules are merely separated from one another. Their mass still exists, as do the atoms themselves.
    b. No, the apparent decrease in matter occurs because gas products escape into the atmosphere. The smoke and gases produced carry some of the mass away from the site of the fire, but the masses of the atoms still exist.

17. Atoms exist with 1, 2, ..., up to 109 protons; each number of protons corresponds to a different element. Most elements exist as isotopes, i.e. atoms with various numbers of neutrons for a given number of protons. The multiple isotopes of each element are the reason for the high number of different mass atoms. For example carbon has three different isotopes; ^{12}C, ^{13}C and ^{14}C.

19. a. ^{50}Ti and ^{50}V are different elements with the same mass number; they are not isotopes.
    b. ^{12}C and ^{14}C are isotopes with same proton count (6 each) but with different neutron counts (6 neutrons in ^{12}C and 8 neutrons in ^{14}C).
    c. ^{40}Ar and ^{40}Kr are not isotopes; Ar and Kr are different elements.

21. Iodine is represented by the symbol $^{127}_{53}I$; it has 53 protons and 74 neutrons. The neutron count equals the mass number minus the atomic number. 127 - 53 = 74 neutrons.

23. In neutral atoms the number of electrons will equal the number of protons. The atomic number equals the number of protons. The mass number equals the sum of the number of protons and the number of neutrons.

	Number of protons	Number of neutrons	Number of electrons	Atomic number	Mass number
a.	32	41	32	32	73
b.	14	14	14	14	28
c.	28	31	28	28	59
d.	48	64	48	48	112
e.	77	115	77	77	192

25. Atomic number or number of protons. This is the quantity that determines the nature of the nucleus. Isotopes have different neutron amounts but the same proton count.

27.

Isotope	Atomic No.	Mass No.	No. of Protons	No. of Neutrons	No. of Electrons
Bromine-81	35	81	35	46	35
Boron-11	5	11	5	6	5
Chlorine-35	17	35	17	18	17
Chromium-52	24	52	24	28	24
Nickel-60	28	60	28	32	28
Strontium-90	38	90	38	52	38
Lead-206	82	206	82	124	82

29. c. The ratio by weight of the elements in the compound.

29. c. The ratio by weight of the elements in a compound. For example samples of methane, CH_4, will always have the relative masses of 12 grams of carbon for every 4 grams of hydrogen.

31. Concepts in science evolve over time and change to account for additional facts.

33. The symbol for krypton is Kr. The atomic number for Kr is 36. The average atomic weight for krypton is 83.80 amu. The electron arrangement has to account for 36 electrons; 2-8-18-8. Much of this information can be taken from the entry for krypton in the periodic table.

36 ← Atomic number
Kr ← Element symbol
83.80 ← Average atomic weight

35.

	Symbol	protons	neutrons	electrons
a.	$^{24}_{12}Mg$	12	12	12
b.	$^{56}_{26}Fe$	26	30	26
c.	$^{115}_{49}In$	49	66	49
d.	$^{127}_{53}I$	53	74	53
e.	$^{107}_{47}Ag$	47	60	47
f.	$^{222}_{86}Rn$	86	136	86

37. Carbon with six electrons has the electron configuration: $1s^22s^22p^2$
   Neon with 10 electrons has the electron configuration: $1s^22s^22p^6$
   Aluminum with 13 electrons has the electron configuration: $1s^22s^22p^63s^23p^1$
   Calcium with 20 electrons has the electron configuration: $1s^22s^22p^6\,3s^23p^64s^2$

39. Na, 1;   Mg, 2;   Al, 3;   Si, 4;   P, 5;   S, 6;   Cl, 7;   Ar, 8

41. Be $1s^22s^2$ B $1s^22s^22p^1$ C $1s^22s^22p^2$ N $1s^22s^22p^3$
   O $1s^22s^22p^4$ F $1s^22s^22p^5$ Ne $1s^22s^22p^6$

43. X-ray(shortest), ultraviolet, visible, infrared, microwave, radio wave(longest)

45. With the electron starting from n=1, the smallest jump is to n=2. Any other jump from the
   n = 1 level would be larger; for example, a jump from n = 1 to n = 3 is a bigger jump than
   from n = 1 to n = 2.

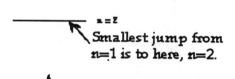

   n=4

   n=3

   n=2
   Smallest jump from
   n=1 is to here, n=2.

   n=1

47. Some lines (wavelengths) observed in the solar emission spectrum did not match
   wavelengths emitted by any of the elements known in 1868. Since each element has its
   own unique emission spectrum, scientists knew they were looking at emissions from a
   previously-undiscovered element.

Solutions for Selected Problems

1.   60.0 g butane x 82.6% C in butane = 60.0 g butane x $\dfrac{82.6 \text{ g C}}{100 \text{ g butane}}$ = 49.6 g C;

   60.0 g butane x 17.4% H in butane = 60.0 g butane x $\dfrac{17.4 \text{ g H}}{100 \text{ g butane}}$ = 10.4 g H

   To check, add the grams of C and grams of H together; the sum should be 60.0 g. Also,
   once either mass has been calculated from the %, it can be subtracted from 60.0 g butane to
   get the mass of the second element in the sample of butane.
   60.0 g butane - 49.6 g C = 17.4 g H or 60.0 g butane - 17.4 g H = 49.6 g C.

2. Electrons all have the same charge. The charge on an oil drop has to be equal to the charge on one or two or three or some whole number of electrons. Let's compare the charges on the oil drops to see if they are related to each other, that is, if one is a factor of all the others. To do this, we choose the smallest oil drop charge (1.33×10^{-19}) and divide each of the others by this same number. If this 1.33×10^{-19} Coulombs is the charge for one electron, then each division will give a quotient that is a whole number or that is very close to a whole number.

$$\frac{1.33 \times 10^{-19}}{1.33 \times 10^{-19}} = 1 \qquad \frac{2.66 \times 10^{-19}}{1.33 \times 10^{-19}} = 2 \qquad \frac{3.33 \times 10^{-19}}{1.33 \times 10^{-19}} = 2.5$$

$$\frac{4.66 \times 10^{-19}}{1.33 \times 10^{-19}} = 3.5 \qquad \frac{7.92 \times 10^{-19}}{1.33 \times 10^{-19}} = 5.95 \text{ or } 6$$

The oil drop charge cannot be a fraction of an electron's charge, such as the charge for 3.5 electrons or 2.5 electrons. The quotients above would each need to be multiplied by 2 to give a set of quotients that are all whole numbers: 2, 4, 5, 7, and 12.

This would mean that the 1.33×10^{-19} charge on the first oil drop is the charge for <u>two</u> electrons, not just one and that the charge for only one electron is

$$\frac{1.33 \times 10^{-19}}{2} = 0.665 \times 10^{-19} = 6.65 \times 10^{-20} \text{ Coulombs.}$$

3. If each taw is 1 inch in diameter (across), a line of 1000 taws would be 1000 inches long. If each taw diameter is measured in centimeters, line of 1000 taws would be 1000 inches x 2.54 cm or 2540 cm long.

This length can be converted meters: $2540 \text{ cm} \times \dfrac{1 \text{ meter}}{100 \text{ cm}} = 25.4 \text{ meters.}$

5. If each marble is 1.0×10^{-8} cm in diameter, then 1000 of them would give a line that is 1000 $\times 1.0 \times 10^{-8}$ cm $= 1.000 \times 10^3 \times 1.0 \times 10^{-8}$ cm $= 1.0 \times 10^{-5}$ cm. This is about 1/10,000 of the thickness of a dime and is too short to measure with a ruler! You would need to line up many more than 1000 marbles to get a length that you can measure.

6. $\dfrac{1.0 \text{ inch}}{1} \times \dfrac{1 \text{ taw}}{1 \times 10^{-8} \text{ cm}} \times \dfrac{2.54 \text{ cm}}{1 \text{ inch}} = 2.54 \times 10^8$ taws or 254,000,000 taws

8. The decimeter equals 0.100 meter.
1 dm is 0.1 m or 1×10^{-1} meters; 1 pm is 1×10^{-12} meters.

Al atoms = $1 \text{ dm} \times \dfrac{1 \times 10^{-1} \text{ m}}{1 \text{ dm}} \times \dfrac{1 \text{ pm}}{1 \times 10^{-12} \text{ m}} \times \dfrac{1 \text{ Al atom}}{286 \text{ pm}} = 3.496 \times 10^8$ Al atoms or

3.50×10^8 Al atoms to 3 significant digit. This is 350,000,000 Al atoms.

11. Measure the distance between two corresponding points on the wave with a ruler. The wave on the left is 1.2 cm long and the wave on the right is 3.9 cm long. This means that their *wavelengths* are 1.2 cm and 3.9 cm. Wavelength, λ, is inversely proportional to energy; this means that the longer the wavelength, the smaller the energy. So, the 1.2 cm wave carries more energy.

Speaking of Chemistry

Name _____

Structure of the Atom

Across

4 The electron _____ for lithium $1s^2 2s^1$.
7 Abbreviation for atomic mass unit.
8 Like charges _____.
11 Proust proposed the law of _____ composition in 1799.
12 Scanning tunneling microscope.
15 Part of atom where mass is concentrated.
17 Particles with +1 charge.
19 John _____ proposed an atomic theory in 1803.
20 Oppositely charged particles_____.
21 Person who proposed nuclear model for the atom.

Down

1 The _____ is determined by the number of outer electrons.
2 Person who proposed a model for the energy states of the electron in hydrogen.
3 _____ charged particles repel
4 Electrode that attracts positive charges.
5 Particles with mass of 1.67×10^{-24} g and no charge.
6 _____ rays have no detectable mass or charge.
9 The electrode that attracts negative particles.
10 Abbreviation for atomic force microscope.
13 Amounts of energy needed for quantum jumps.
14 _____ number indicates the number of protons in a nucleus.
16 Particles identical with electrons.
18 Lowest energy state for an atom.

Bridging the Gap   I
Seeing the Light:
Incandescent, Mercury Vapor or Sodium Vapor.

Street lighting and lighting of parking lots is a common public service. The type of electric street lamps used has changed over the years. Incandescent lamps that depended on glowing white hot filaments for light were replaced in the 1950s. Currently, street lamps are mercury vapor lamps, sodium vapor lamps or metal halide lamps (these contain thallium, indium, mercury and sodium). These are cheaper to operate than incandescent lamps because they use less electricity to produce the desired light intensity (lumens per watt). The colors emitted by these lamps are very different. The mercury vapor lamp produces a bluish light. A sodium lamp produces a yellowish color and the metal halide lamps are almost white. The energy carried by a photon of blue light is different from the energy carried by a photon of yellow light. The energy needed to produce these two types of light is also different. Check the text Figure 3.15 on page 70 which shows the continuous spectrum of light and determine which of these two colors carries more energy per photon.

This exercise has four parts. One part is to decide which type of lamp requires less energy. Part two is to determine which kind of street lighting is used in your community by observing the colors of the street lights. Are your street lights mercury vapor lamps, sodium vapor lamps, incandescent lamps, metal halide lamps or are they an assortment of all kinds? A third part is to write a brief argument for standardized type of street lamp. Lastly you are to write one question you would want answered by an expert before you would make a final choice of street lamp.

Bridging the Gap   I
Concept Report Sheet
Name _____

Seeing the Light:

Incandescent, Mercury Vapor or Sodium Vapor.

1.   Type of lamp that uses less energy (include a supporting reason for your answer)

2.   Type of street lamp that is most common in your community
     (Tell how you came to this conclusion.)

3.   Argument for having a single type of street lamp

4.   The question you want an expert to answer so you can make a better choice.

Bridging the Gap II   Name _____
Radiation sources: cathode ray tubes, light bulbs, microwaves, remote controls, hair dryers and toasters

This exercise is aimed at heightening your awareness of radiation sources. These devices are an outgrowth of the experiments described in Chapter 3. You are supposed to do an inventory of your surroundings and count the number of electromagnetic sources that you encounter each day (stop at 3 of each type). You will be surprised at how many sources exist around you. Pretend you work for the American Consumer Safety Group. What question would you want answered about these sources and the radiation they emit.

Cathode ray tubes are the heart of today's television sets. The continuous spectrum is emitted by most incandescent light bulbs; heating elements in toasters, hair dryers, clothes dryers, ovens and electric stove elements all glow red and emit light. The light emitting diodes on the digital displays of clock radios, televisions, etc. all emit specific forms of light. The remote control for your electronic equipment emits infrared light. If you have a garage door opener or a remote control for a car alarm, you are using a radiation source.

Concept Report Sheet

1.   Locations of cathode ray tubes

   _____

2.   Locations of visible light sources other thanlight bulbs.

   _____

   Locations of light bulbs

   _____

3.   Locations of red light, microwave and infrared emitters

   _____

4   Question you want answered regarding these radiation sources.

Solution to Speaking of Chemistry crossword puzzle

Structure of the Atom

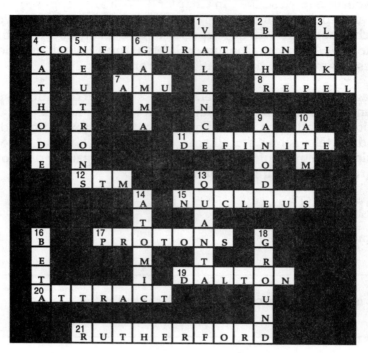

4   The Periodic Table

4.1   Development of the Periodic Table

This section describes Mendeleev's and Meyer's work and their proposal of the periodic law and a periodic arrangement of the elements based on atomic mass. Mosley's experiments that linked periodic properties to atomic number are also discussed.

Objectives

After studying this section a student should be able to:

give the definitions for atomic mass and atomic number
describe Mendeleev's contribution to the development of the periodic table
describe the historical development of the periodic table and state the periodic law
identify the atomic property that governs the position of an element in the periodic table

Key terms

atomic weights	trend in properties	periodic law
increasing atomic number	Mendeleev	Meyer
repeatable trend	periodic trend	physical properties

4.2   The Modern Periodic Table

This section gives the reasons for the classification of elements into groups. The definition of a period is given and explained. It describes the physical properties that are used to classify elements as metals, nonmetals, and metalloids. Examples of elements in these classifications are given. The groups for semi-conductors and noble gases are identified.

Objectives

After studying this section a student should be able to:

describe the significance of a group in the periodic table
locate the main-group, transition and innertransition elements
tell how main-group elements differ from the rest of the periodic table
tell the group of an element if given the element symbol and a periodic table
describe the physical properties of metals, nonmetals and metalloids
explain why the periods of the main group elements are typically eight elements long
locate metals, nonmetals, noble gases and metalloids on the periodic table

Key terms

representative groups	main-group	semiconductors
B groups	transition elements	innertransition elements
lanthanide series	actinide series	periods
metals	malleability	ductility
conduction of heat	conduction of electricity	positive ions
nonmetals	insulators	metalloids

4.3   The Periodic Table and Chemical Behavior

This section describes the Lewis dot symbol notation for atoms and ions. It shows how the Lewis dot symbols for atoms change when the atoms lose electrons to form cations and how the symbols change when the atoms gain electrons to form anions. The section includes an explanation of the importance of valence electrons in chemistry.

Objectives

After studying this section a student should be able to:

   give the definition for valence electrons
   write the Lewis dot symbol for a representative element if given a periodic table
   give the definition for a Lewis dot symbol
   tell how group number and valence electrons relate
   write the Lewis dot symbols for the reaction of a metal being converted to a cation
   write the Lewis dot symbols for the reaction of a nonmetal being converted to an anion
   explain why valence electrons are chemically interesting

Key terms

kernel	outermost shell	reactivity
valence electrons	Lewis symbol	dot symbol

4.4   Periodic Trends in Atomic Properties

This section describes how metallic properties, atomic radii, and reactivity vary for elements in the same group. The variations of these properties for elements in a period are described and explained. The relationship between atomic size and reactivity is explained for metals and nonmetals.

Objectives

After studying this section a student should be able to:

   describe and explain how atomic radii vary for the members of a group
   describe and explain the variation of atomic radii across a period
   describe and explain how reactivity varies for the members of a group
   relate reactivity to atomic radius

Key terms

metallic character	atomic properties	atomic radii (radius)
increase in reactivity	gain electrons	lose electrons

4.5   Properties of Main-Group Elements

This section includes descriptions of the physical properties and the uses of main group elements and their compounds. Formulas for oxides and important compounds formed by the representative elements are given. Sources and specific applications of the alkali metals, alkaline earths, Group IIIA and IVA elements and halogens are described. The xerographic process is outlined. Sir William Ramsay's Nobel Prize work to isolate the noble gases is outlined.

Objectives

After studying this section a student should be able to:

   identify the representative elements using a periodic table
   tell which elements are metals, nonmetals and metalloids
   match elements with their group: halogens, noble gases, alkaline earth metals, etc.
   give an example of a compound formed by a typical member of each group
   tell what is wrong with calling the noble gases "inert"
   give the formula for the oxide compound formed by a main group element i.e. M_xO_y
   give the formula for the halide compound formed by a main group element i.e. M_xX_y
   summarize the work done by Ramsay to isolate the noble gases

describe the general properties of members of each group

name a common substance formed by an element from each group

tell how an alloy is made

Key terms

group properties	alkali metals	alkaline earth metals
halogens	noble gases	distillation of liquid air
Ramsay	liquid oxygen	alloys
salts	xerography	fertilizers
explosives	oxide	electrostatic

4.6   Properties of Transition Group Elements

This section gives the reasons for similar behavior of transition elements. The differences between 4th period elements and the 5th and 6th period elements are explained.

Objectives

After studying this section a student should be able to:

locate the transition group elements in the periodic table

explain why 5th and 6th period elements are different from 4th period elements

describe how atomic radius varies across the 4th, 5th and 6th periods

name three transition group elements that exist naturally as pure elements

Key terms

transition group elements	metals	period 4
period 5	period 6	free element

Additional readings

Everhart, Thomas E. & Thomas L. Hayes. "The Scanning Electron Microscope" <u>Scientific American</u> Jan. 1972: 54.

Fowler, William A. "The Origin Of The Elements" <u>Scientific American</u> Sept. 1956 : 82.

Letokhov, Vladilen S. "Detecting Individual Atoms And Molecules With Lasers" <u>Scientific American</u> Sept. 1988: 54 .

Muller, Erwin W. "Atoms Visualized" <u>Scientific American</u> June 1957: 113.

Perlman, I. & G. T. Seaborg. "The Synthetic Elements" <u>Scientific American</u> April 1950: 38.

Zare, Richard N. "Laser Separation Of Isotopes" <u>Scientific American</u> Feb. 1977: 86.

Answers to Odd Numbered Questions for Review and Thought

1. When elements are arranged in order of their Atomic Numbers, their chemical and physical properties show repeatable trends.

3. Metals are elements that show metallic physical properties such as malleability, ductility, good electrical and thermal conductivity. Malleability means a metal can be hammered into sheets. Ductility means a metal can be drawn into wires. Metals are used for electrical wires and cooking pots because of their conductivity. Chemically, metals react to lose electrons and form positive ions. Metals are the elements, except for hydrogen, lying to the left and below the diagonal line formed by the metalloids in the Periodic Table.

metals	
metalloids	
nonmetals	

1 IA	2											13		14	15	16	17	18 VIIA
H	IIA											IIIA	IVA	VA	VIA	VIIA	He	
Li	Be	3 IIIB	4 IVB	5 VB	6 VIB	7 VIIB	8 VIIIB	9 VIIIB	10 VIIIB	11 IB	12 IIB	B	C	N	O	F	Ne	
Na	Mg											Al	Si	P	S	Cl	Ar	
K	Ca	Sc	Ti	V	Cr	Mn	Fe	Co	Ni	Cu	Zn	Ga	Ge	As	Se	Br	Kr	
Rb	Sr	Y	Zr	Nb	Mo	Tc	Ru	Rh	Pd	Ag	Cd	In	Sn	Sb	Te	I	Xe	
Cs	Ba	La	Hf	Ta	W	Re	Os	Ir	Pt	Au	Hg	Tl	Pb	Bi	Po	At	Rn	
Fr	Ra	Ac	†															

† The remaining elements are metals.

5. Metalloids are elements like boron, B, and germanium, Ge, that have properties intermediate between those of metals and nonmetals. They will conduct electricity under special conditions and can lose electrons to form positive ions.

7. The answer to this question depends on using the periodic table inside the back cover of the text book. The following pairs of elements would be incorrectly placed in the periodic table if atomic mass were used to sequence the elements instead of atomic numbers: 27 and 28, cobalt and nickel; 52 and 53, tellurium and iodine; 90 and 91, thorium and protactinium; 92 and 93, uranium and neptunium; 94 and 95, plutonium and americium; 96 and 97 are equal, curium and berlklium; 106 and 107, seaborgium and nielsbohrium

9. Metals: Na, sodium; Pb, lead. Nonmetals: H, hydrogen; Cl, chlorine. Metalloids: As, arsenic; Ge, germanium.

11. Metals lose electrons to form positive ions, are malleable and ductile, are good conductors of heat and electricity. Nonmetals gain electrons to form negative ions, are brittle, are insulators not conductors.

13. a. The periods are the horizontal rows in the periodic table. There are 7.
    b. There are 8 representative groups. These are identified by a Roman numeral and a capital "A". The first representative group is IA. It includes H, Li, Na, K, Rb, Cs, and Fr. The last representative group is VIIIA. It includes He, Ne, Ar, Kr, Xe, Rn. These are the rare or noble gases.
    c. There are two representative groups that are all metals if you ignore H in IA. Group IA includes Li, Na, K, Rb, Cs, and Fr. The other is IIA. It includes Be, Mg, Ca, Sr, Ba, and Ra.
    d. The elements in Group VIIA and VIIIA are all nonmetals.

e. Yes, all the elements in period 7 are metals. These are
   Fr   Ra   Ac   Th   Pa   U   Np   Pu   Fm   Cm   Bk   Cf
   Es   Fm   Md   No   Lr   Rf   Ha   Sg   Ns   Hs   Mt

15. All elements in the same group have the same number of valence electrons and have similar chemical properties. For example all of the elements in Group IA have one valence electron and react to form positive ions with a plus one charge. Lithium loses one electron to make a Li^{1+} ion. Nonmetals in the same group will usually form the same number of

   covalent bonds. The Group IVA elements typically form four bonds such as CH_4 and

   SiF_4, . Similarly elements in Group IIIA form three bonds, BF_3,

17. Atomic radius is the distance from the center of the nucleus to the outer surface of the electron cloud around the nucleus.
    Reactivity is a measure of how readily the substance enters into a chemical reaction. The element helium is so unreactive it doesn't combine with any other element to form a compound. Oxygen is extremely reactive and combines readily with other elements to form oxides such as CO_2 and Fe_2O_3.

19. All of the alkali metals Li, Na, K, Rb, Cs, and Fr have one valence electron.

21. a   Barium, Ba, is in Group IIA and has 2 valence electrons.
    b. Aluminum, Al, is in Group IIIA and has 3 valence electrons.
    c. Phosphorus, P, is in Group VA and has 5 valence electrons.
    d. Selenium, Se, is in Group VIA and has 6 valence electrons.
    e. Bromine, Br, is in Group VIIA and has 7 valence electrons.
    f. Potassium, K, is in Group IA and has 1 valence electron.

23. The Lewis symbol can be determined by locating the element in the periodic table and identifying the group number. For example, Beryllium is in Group IIA. It has two valence electrons and two dots in the Lewis symbol. Similarly Potassium has one valence electron

   and one dot.   •Be•   K•   :C̈l•   •As•   :Kr:

25. a. Li is more metallic than F, fluorine. When elements are in the same row or period, the ones to the right in the periodic table are less metallic.
    b. Cs is more metallic than lithium. When elements are in the same group the elements
    at   the bottom of a group are more metallic. They have larger atoms and lose electrons more readily.
    c. Ba   Same reason as 25 b above.
    d. Pb   Same reason as 25 b above.
    e. Al Same reason as 25 b above.
    f. Na   Same reason as 25 a above.

27. The noble gases have 8 valence electrons. This is the reason that they usually do not combine with other elements. The noble gases have a set of electrons that satisfies the octet rule.

29.

Atomic number	Element name	Number of valence electrons	Period	Metal, M or Nonmetal, NM
6	carbon, C	4	2	NM
12	magnesium, Mg	2	3	M
17	chlorine, Cl	7	3	NM
37	rubidium, Rb	1	5	M
42	molybdenum, Mb	6	5	M
54	xenon, Xe	8 or 0	5	NM

31.   The smallest member of a group typically is at the top of the group. F < Cl < Br < I < At

33.   Beryllium, Be   Magnesium, Mg   Calcium, Ca
       Strontium, Sr   Barium, Ba   Radium, Ra

35.   The smaller the radius of the nonmetal atom, the more reactive it is. This is reasonable
       because the nonmetals are short of an octet. The atomic nucleus will be have a stronger
       attraction for an electron from another atom because there are fewer electrons
       surrounding the nucleus.

37.   Both atomic size and metallic character decrease from left to right across a period; both
       atomic size and metallic tendency increase from top to bottom of a group.
       Relative atomic radii for period 2   Relative atomic radii for Group IA

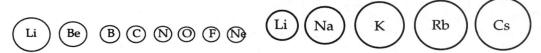

39.   There is a stepwise change in proton number across a period. The attraction between the
       nucleus and the electrons in the atom gradually increases because of this change. This is
       duplicated in each period so trends in properties appear for each period.

41.   K (largest), Al, P, S, Cl (smallest)

43.   The elements in Group IVA change from nonmetal to metal. Carbon at the top of the
       group is not malleable, it does not conduct electricity well, and it does not lose electrons
       to form positive ions. Carbon is a nonmetal with nonmetallic properties. Lead, Pb, is a
       metal. It is malleable and can be hammered into sheets or drawn into wire. It loses
       electrons to form positive ions.

45.   Helium: cooling gas, balloons and as an additive to pressurized oxygen breathing gases to
       prevent the occurrence of the bends. Neon and Argon: lighting, electric discharge tubes
       and an inert environment for welding.

47. Compounds of xenon, krypton and radon have been produced so all group members are not totally unreactive or inert. the formulas for these compounds do not follow the octet

rule. Example noble gas compounds are KrF_2, and XeF_4

49. The number of oxygen atoms per one atom of the other element increases by 1/2 from one group to the next: MgO (1 to 1), Al_2O_3 (1 to 1.5), SiO_2 (1 to 2), etc. This leads to a predicted formula of Cl_2O_7. The progression is summarized in this table.

1 to 1/2	1 to 1	1 to 3/2	1 to 4/2	1 to 5/2	1 to 6/2	1 to 7/2
Na_2O	MgO	Al_2O_3	SiO_2	P_2O_5	SO_3	Cl_2O_7

51. Both oxygen and sulfur have 6 outer electrons. Both form negative ions by gaining two electrons. The ions are oxide, O^{2-}, and sulfide, S^{2-}. The ions are formed when the atoms gain two electrons to complete their octets. The sulfur and oxygen atoms usually form two covalent bonds. The hydrides are H_2O and H_2S.

53. False; the majority of elements are metals. There are only 17 nonmetal elements.

55. Atomic weight for Na is 22.99 amu. Density is 0.7 g/cm^3. The melting point is 122°C and the boiling point is 1044°C

57. Element 119 would fall under Fr, francium. It would be in Group IA and be the first element of period 8.

Speaking of Chemistry

Name _____

Periodic Trends

Across

2 Nonmetals react to _____ electrons.
4 Horizontal rows of elements.
5 Elements in Group VIII A.
7 Mixture of metals.
10 Describes elements in the A groups.
15 _____ dot formulas indicate number of valence electrons.
17 Elements that conduct heat and lose electrons.
18 Radioactive noble gas.
19 Element in Group IA.
21 Noble gas with smallest radius.

Down

1 Element from Group IIIA.
3 Name for Group IA elements.
6 Number of valence electrons in Na.
8 Father of the periodic table.
9 Vertical arrangements of elements that have similar properties.
11 Oxygen has ___ valence electrons.
12 Elements that tend to gain electrons.
13 Number of valence electrons in Al.
14 Another name for outer electrons.
16 Rare element used in Xerography.
20 Many names for metals end in ____

Bridging the Gap
Concept Report Sheet
Predicting Properties of Elements

Name _____

The strength of any theory lies in its usefulness to make predictions about new and untried situations. This exercise is aimed at demonstrating the power of the periodic law. The text points out on page 96 that the periodic table is being extended by synthesizing new elements using high energy particles in nuclear accelerators. Element 118 is projected to be a noble gas. This exercise is intended to use the periodic law and trends in the table to predict the properties for a different element, number 119. Information given in the chapter should allow you to predict values and properties for element 119. Write your answers on this sheet.

1. Prediction of Group and Lewis symbol
    What group of representative elements would include atomic number 119 ? Write the Lewis symbol for element 119.(Use Uk for the element symbol.) Explain your reasoning for picking this group. Pick a name and symbol for this element. Tell why you made this choice of a name and symbol.

2. Prediction of atomic properties: approximate atomic weight and atomic radius
    What is the likely atomic weight for atomic number 119? Explain how you made your prediction.

    What is the likely atomic radius? Explain how you made your prediction.

3. Prediction of physical properties: melting point, boiling point and density
    Give values for the following and describe how you made your choices.

density	boiling point	density

4. Prediction of chemical properties: classification as a metal, nonmetal or metalloid;
    reactivity with water; and formula for chloride
    What classification would you assign to this element? Explain your reasoning.

    Do you predict that element 119 will react with water? Justify your answer.

    What formula do expect for the combination of chlorine and element 119? Write the
    equation for the reaction between chlorine, Cl_2, and element 119.

52

Solution to Speaking of Chemistry crossword puzzle

Periodic Trends

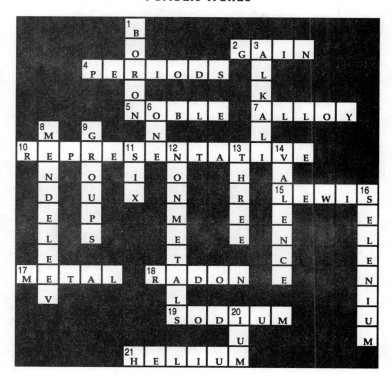

5 Nuclear Changes

5.1 The Discovery of Radioactivity

Becquerel's discovery of radioactivity is described. The effect of radioactivity from uranium on electroscopes is given as proof that it is a form of ionizing radiation. The studies by Rutherford and Villard on the penetrating power of radiation are summarized.

Objectives

After studying this section a student should be able to:

describe Becquerel's experiment that proved that phosphorescence and radioactivity were different phenomena

describe the mass and charge of alpha and beta particles

tell how the penetrating powers of alpha, beta and gamma rays differ

explain what is meant by the term ionizing radiation

Key terms

X rays	phosphoresce	emitted radiation
ionize	vacuum jar	electroscope
alpha, beta, gamma rays	penetrating ability	

5.2 Nuclear Reactions

The differences between chemical reactions and nuclear reactions is summarized. The method for Balancing nuclear reactions is described and illustrated for the alpha and beta emissions.

Beta particle formation process is explained and illustrated. $^{1}_{0}n \rightarrow ^{0}_{-1}e + ^{1}_{1}H$

Beta emission

$$^{210}_{82}Pb \rightarrow ^{210}_{83}Bi + ^{0}_{-1}e$$

Alpha emission

$$^{218}_{84}Po \rightarrow ^{214}_{82}Pb + ^{4}_{2}He$$

Objectives

After studying this section a student should be able to:

state the definition for radioactivity, nucleons and transmutation

describe what a nuclear reaction is and give an example

tell how a nuclear reaction differs from a chemical reaction

balance a nuclear equation if given the starting nucleon and the type of emission

identify a balanced nuclear equation

tell how nuclear charge and mass change when either an alpha or beta particle is emitted

Key terms

nuclear reaction	transmutation	nucleons
mass number	nuclear equation	beta-emitter
alpha-emitter	$^{4}_{2}He$	$^{0}_{-1}e$

5.3 The Stability of Atomic Nuclei

The stability of nucleons is explained in terms of the plot of number of neutron versus number of protons for known isotopes. The band of stability is described and types of emissions are explained in terms of the position of a nuclide relative to the band of stability. Positron emission is illustrated.

Objectives

After studying this section a student should be able to:

   give the standard symbol for atomic number

   describe the properties of a positron

   tell how nuclear instability is linked to atomic number 83

   tell how the atomic number and the mass number change for an isotope after an event
      such as alpha emission, beta emission, positron emission

   tell how neutron number and atomic number are related in stable nuclei

   predict the kind of particle emitted by an atom with too few neutrons i.e. $^{13}_{7}N$

   predict the kind of particle emitted by an atom with too many neutrons i.e. $^{210}_{82}Pb$

   tell what type of emission is expected for atoms that are too big such as $^{230}_{90}Th$

Key terms

stability of nuclei	anti-electron	positron emission
unstable isotope	too many neutrons	too few neutrons
positron	decay	$^{0}_{+1}e$

5.4   Activity and Rates of Nuclear Disintegrations

This section defines half life. Examples describe how radioactive source activity and mass change with time. Graphs of activity versus half life are explained. The curie and bequerel are defined. The natural decay series are discussed and the specific steps in the uranium-238 decay series are graphed.

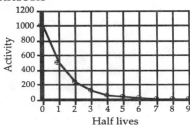

Objectives

After studying this section a student should be able to:

   give the definition for a Bequerel, half life, and activity in terms of radioactivity

   tell what a curie (Ci) equals in terms of disintegrations per second

   state what a microcurie (μCi) equals in terms of disintegrations per second

   predict the activity of a source if given initial activity, elapsed time and the half life period

   graph the mass of radioactive material versus time if given initial mass and half life period

   tell what is meant by a natural radioactive decay series

Key terms

activity	highly radioactive	radioactive disintegrations
microcurie, μCi	curie, Ci	Becquerel, Bq
half-life	cobalt-60	gamma emitter
decay series	uranium series	thorium series
actinium series		

5.5   Atomic Dating

This section describes how the age of rocks geologic formations can be determined using the natural decay series for uranium-238 to lead-206 . The carbon-14 dating method is explained. The activity of carbon-14 per gram of carbon is explained. The process for replacing carbon-14 in the atmosphere is outlined and the reason why carbon-14 levels stay constant in living organisms is summarized.

Objectives

After studying this section a student should be able to:

   describe how the ratio of the mass of uranium-238 to mass of lead-206 compare after one
      half life, two half lives, three half lives, etc.

   describe what is meant by a rate determining step

   tell why the level of carbon-14 stays constant in living tissue

   explain why the carbon-14 levels in decrease after the organism dies

   estimate the age of a radioactive source if given the half life period, initial activity or mass
      and present activity or mass

   describe the contributions of Rutherford, Berquerel, Curie and Villard

   state what the carbon-14 activity level is in disintegrations per second per gram of carbon

   read a graph of activity versus time and estimate the activity after some elapsed time

Key terms

   rate-determining-step   atomic dating   carbon-14 dating

5.6   Artificial Nuclear Reactions

This section describes the nuclear bombardment process, summarizes the historical
development of artificial nuclear transformation, explains why nuclear bombardment and
cyclotrons are used in medicine, shows in figure 5.7 how simple bombardment events occur.

Objectives

After studying this section a student should be able to:

   describe what is meant by a compound nucleus

   describe how Rutherford produced the first artificial nuclear change

   give the definition for transmutation

   describe the roles of Irene Curie Joliot and Frederic Joliot in making artificial radioisotopes

   explain why hospitals and medical research labs often operate cyclotrons

   describe how cyclotrons are used in bombardment processes

Key terms and equations

   artificial nuclear reaction   artificial transmutation   compound nucleus
   kinetic energy   bombardment reaction   cyclotron

5.7   Transuranium Elements

This section describes the transuranium elements, explains how the artificial transuranium
elements beyond $Z = 92$ are produced, gives example decay reactions for transuranium
elements, and tells why transuranium elements are made. Nuclear fission is described.
Uranium enrichment is explained.

Objectives

After studying this section a student should be able to:

   give the definition of a transuranium element

   identify a transuranium element if given its symbol or atomic number

   tell what kinds of projectile particles are used to produce elements up to $Z = 101$

   balance a nuclear bombardment equation and identify the product nucleus given the target
      and bombardment particle

   describe the changes in the periodic table that were proposed by Glenn T. Seaborg

   describe a fission reaction

   tell what is meant by fissionable material and explain what enriched uranium means

Key terms

transuranium elements   fissionable material   fission
plutonium   uranium   bombardment experiment

5.8 Radiation Effects

This section describes the principal factors determining the hazards a radioactive substance poses: activity, energy or type of radiation, role in food chain or body functions. The röntgen (R), the rem and the millirem are defined and described. Whole-body radiation effects, somatic and genetic, are defined and explained.

Objectives

After studying this section a student should be able to:

state factors that determine the hazard posed by a radioactive substance
tell how a dental x-ray and the röntgen are related
describe how a somatic effect differs from a genetic effect
explain how genetic effects result from radiation exposures
tell what is meant by an induction period

Key terms

nuclear reaction	cosmic rays	X rays
radiation damage	normal cell processes	disintegrations per second
type of radiation	energy of radiation	incorporated into food chain
dose	röntgen, R	rem
millirem, mrem	whole-body radiation	somatic effects
genetic effects	induction period	mutation

5.9 Radon and Other Sources of Background Radiation

This section describes background radiation, from both man made and natural sources. It gives a detailed discussion of radon, the dominant natural source of ionizing radiation. The half life of radon-222 and the hazards posed by its daughters are described. The methods used to mitigate "action level" radon concentrations are explained.

Objectives

After studying this section a student should be able to:

name an example of a natural and a man-made radiation source
tell how much of the average radiation exposure comes from man-made sources and how
    much comes from radon and other natural sources
describe the radon decay reaction and explain why it is a lung cancer hazard
name the radon daughters and the type of emissions released in their decay processes
explain how background radiation arises
tell which regions of the United States have the greatest radon potential

Key terms

ionizing radiation	background radiation	radon daughters
epithelial cells	chimney effects	DNA
risk assessment models	EPA	picocurie, pCi

5.10 Useful Applications of Radioactivity

This section reviews some beneficial uses of radioisotopes such as food preservation, materials testing, tracers for following material pathways, medical imaging and diagnosis. The problems associated with these applications are described.

Objectives
After studying this section a student should be able to:

- describe how foods can be preserved using gamma rays from sources like cobalt-60
- give a definition of "radiolytic products"
- explain how cobalt-60 gamma radiation is used in materials testing
- tell how phosphorus-32 is used in tracer studies
- name a diagnostic radioisotope and tell what organs it is used to image
- describe how positron emission tomography (PET) produces detectable gamma rays
- tell how salmonella contamination can be eliminated by irradiation

Key terms

gamma rays
irradiated food
poultry irradiation
tagging
metastable
PET
annual dose

food irradiation
radiolytic products
materials testing
nuclear medicine
medical imaging
PET scanner

FDA
Salmonella
radioactive tracers
diagnostic radioisotopes
positron emission tomography
antimatter-matter annihilation

Additional readings

Armbruster, Peter & Gottfried Munzenberg. "Creating Superheavy Elements" Scientific American May 1989: 66.

Blosser, Henry G. "Medical Cyclotrons" (Special Issue: Physical Review Centenary - From Basic Research to High Technology) Physics Today Oct. 1993: 70.

Damon, P.E., et al. "Radiocarbon Dating and the Shroud of Turin" Nature Feb. 1989: 611.

Deevey, Edward S. , Jr. "Radiocarbon Dating" Scientific American Feb. 1952: 24

Ghiorso, Albert & Glenn T. Seaborg. "The Newest Synthetic Elements" Scientific American Dec. 1956: 66.

Hamilton, J. H. & J. A. Maruhn. "Exotic Atomic Nuclei" Scientific American July, 1986: 80

Farmelo, Graham. "The Discovery of X-rays" Scientific American Nov. 1995: 86.

Litke, Alan M. & Richard Wilson. "Electron-positron Collisions" Scientific American Oct. 1973: 104

Neher, Paul "Radon Monitor" Electronics Now Feb. 1994: 66.

Platzman, Robert L. "What Is Ionizing Radiation?" Scientific American Sept. 1959: 74.

"Radon: Worth Learning about." (Includes an article on home radon detectors) (Special Report: Safe at Home) Consumer Reports July 1995: 464(2).

Spoerl, Edward. "The Lethal Effects Of Radiation" Scientific American Dec. 1951: 22.

Weaver, Kenneth F. "The Mystery of the Shroud" National Geographic June 1980: 730.

Answers to Odd Numbered Questions for Review and Thought

1. Gamma rays can pass through more than 22 cm of steel and about 2.5 cm of lead. Alpha particles can be stopped by a thin piece of paper or 2 to 8 cm of air. Beta particles have greater penetrating power.

3. The more hazardous radioisotope is $^{222}_{86}Rn$ with a short half-life of 3 days because more ionizing radiation is emitted over a shorter time period. This means anyone near the source during this time will be exposed to a higher number of emitted particles and a greater risk.

5. An annihilation reaction occurs. The two particles are converted to energy in the form of gamma radiation. see page 140. $^{0}_{1}e + ^{0}_{-1}e \longrightarrow 2\ ^{0}_{0}\gamma$

7. The law of conservation of matter requires the mass numbers of the products equal the sum of the mass numbers of reactants.

9. The production of the artificial elements up to atomic number 101 were made using $^{4}_{2}He$ or $^{1}_{0}n$ projectiles; for example Californium can be produced by the following process:

$$^{4}_{2}He + ^{239}_{94}Pu \longrightarrow ^{240}_{96}Cm + ^{1}_{0}n$$

11. Transuranium elements have atomic numbers greater than 92 and are all artificial. None of the transuranium elements are found in nature.

13. The "actinium series" is a decay series that starts with U-235 and ends with Pb-207. The series consists of a number of decay steps. An example of one set of steps is shown below. The $\xrightarrow{\alpha}$ indicates an alpha decay step and the $\xrightarrow{\beta}$ indicates a beta decay step.

$$^{235}_{92}U \xrightarrow{\alpha} {}^{231}_{90}Th \xrightarrow{\beta} {}^{231}_{91}Pa \xrightarrow{\alpha} {}^{227}_{89}Ac \xrightarrow{\alpha} {}^{223}_{87}Fr \xrightarrow{\beta} {}^{223}_{88}Ra \xrightarrow{\alpha} {}^{219}_{86}Em \xrightarrow{\alpha}$$
$$^{215}_{84}Po \xrightarrow{\beta} {}^{215}_{85}At \xrightarrow{\alpha} {}^{211}_{83}Bi \xrightarrow{\beta} {}^{211}_{84}Po \xrightarrow{\alpha} {}^{207}_{82}Pb$$

15. Gamma radiation of food prevents spoilage. Living organisms such as bacteria, molds, and yeasts are killed by the gamma radiation. Growth and maturation processes are inhibited in fresh foods. This prolongs shelf life of the food fresh food.

17. Cyclotrons accelerate projectile particles to high enough velocities so collisions between projectile and target nuclei result in fusion of the nucleons, forming a new nucleus. Neutron deficient radioisotopes such as carbon-11 or oxygen-15 are produced in the cyclotron. These are positron emitters with short half-lives. The positrons are detected when they collide with an electron and both particles are annihilated. See question 5.

19. The gamma rays can damage DNA sequences necessary for cell reproduction, thereby slowing or stopping cell reproduction. Cancer cells reproduce faster than normal cells and are more sensitive to this because the DNA sequence is needed for duplication of the genetic code more often in the rapidly multiplying cancer cells.

21. Natural radiation in food, water and air exceeds the exposure from weapons fallout. The radiation coming from food, water, and air equals 24 mrem. The exposure from weapons fallout is 4 mrem per year.

23. Gamma irradiation would kill the E. coli. This would eliminate the possibility of E. coli poisonings. Public concern about potential risks from radiolytic by products prevent the adoption of radiation sterilization. There is a willingness to accept the risk of E. coli poisonings that accompany the present methods of handling food.

25. This indicated that the radioactive materials gave off ionizing radiation that exposed the film.

27. a. In 1899 Rutherford discovered that alpha rays could be stopped by thin layers of paper and that beta rays were more penetrating.
    b. In 1934 Irene and Frederic Joliot made the first artificial radioactive isotope, phosphorus-30.
    c. The first atomic bomb explosion occurred in 1945.
    d. Becquerel discovered that uranium emitted ionizing radiation in 1896.
    e. Additional nuclear testing occurred in 1946.

29. The emission of a beta particle has no effect on the mass number of the daughter because the mass number for a beta particle is zero.

$$^{211}_{82}Pb \rightarrow ^{211}_{83}Bi + ^{0}_{-1}e$$

Solutions for Selected Problems

1. a. $^{64}_{29}Cu \longrightarrow ^{64}_{30}Zn + ^{0}_{-1}e$ b. $^{69}_{30}Zn \longrightarrow ^{69}_{31}Ga + ^{0}_{-1}e$ c. $^{131}_{53}I \longrightarrow ^{131}_{54}Xe + ^{0}_{-1}e$

3. a. $^{222}_{86}Rn \longrightarrow ^{4}_{2}He + ^{218}_{84}Po$ b. $^{225}_{90}Th \longrightarrow ^{4}_{2}He + ^{221}_{88}Ra$

4. The number of atoms will decline by 1/2 after each half-life. There will be 150000 atoms after one half-life; 75000 atoms after two half-lives; 37500 atoms after three and 18750 atoms after four. The graph illustrates the decrease in the number of atoms.

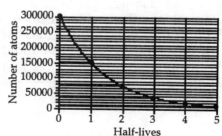

Alternatively, the number of atoms remaining after "n" half-lives can be calculated from the equation: amount remaining $= (\text{initial amount})\left(\dfrac{1}{2}\right)^n$

amount remaining $= (300,000 \text{ atoms})\left(\dfrac{1}{2}\right)^4 =$

$(300,000 \text{ atoms})\left(\dfrac{1}{2}\right)\left(\dfrac{1}{2}\right)\left(\dfrac{1}{2}\right)\left(\dfrac{1}{2}\right) = (300,000 \text{ atoms})\left(\dfrac{1}{16}\right) = 18750 \text{ atoms}$

5. Activity remaining $= \left(\text{initial activity}\right)\left(\dfrac{1}{2}\right)^{n}$ where n = number of half-lives that elapse.

activity remaining $= \left(40{,}000\ \text{Bq}\right)\left(\dfrac{1}{2}\right)^{3} = \left(40{,}000\ \text{Bq}\right)\left(\dfrac{1}{2}\right)\left(\dfrac{1}{2}\right)\left(\dfrac{1}{2}\right) =$

$\left(40{,}000\ \text{Bq}\right)\left(\dfrac{1}{8}\right) = 5{,}000\ \text{Bq}$ where 1 Bq = 1 disintegration per second.

9.

cosmic radiation at sea level	26
U.S. elevation at San Diego	0
ground	26
wood house	0
food, water, air average	24
weapons test fallout	4
chest X-ray 2 at 10 each	20
jet plane travel	0
TV viewing 3 hours at 0.15 per hour	.45
nuclear plant 100 miles away	0
total	100.45

This 100.45 mrem is below the U.S. annual average. The greatest contribution to this person's exposure comes from where he/she lives--the 26 mrems from cosmic radiation and the 26 mrems from ground radiation.

11. Alpha particles can be stopped by a few sheets of paper. Beta particles can penetrate paper but can be stopped by a thin (0.5 cm) sheet of lead. Gamma radiation would pass through the paper and the thin sheet of lead and would require 2.5 cm of lead to stop it.

12. $\left(55\ \text{kg body mass}\right)\left(\dfrac{1000\ \text{g body mass}}{1\ \text{kg body mass}}\right)\left(\dfrac{0.004\ \text{g C}}{100\ \text{g body mass}}\right)\left(\dfrac{15.3\ \text{disintegrations / minute}}{1\ \text{g C}}\right)$

= 3366 disintegrations/minute; this solution method is based on conversion factors. Alternatively, the problem can be worked "in pieces": change the 55 kg body mass to 55000 grams, then find the mass of carbon in the body by finding 0.4% of 55000 g = 220 g C. If there are 15.3 disintegrations/minute per 1 gram of C, then for 220 g C there will be:

$\left(220\ \text{g C}\right)\left(\dfrac{15.3\ \text{disintegrations / minute}}{1\ \text{g C}}\right) = 3366$ disintegrations/minute.

62

Speaking of Chemistry

Name _____

Applications of Radioactivity

Across

8 Changing one element into another.
9 Food and Drug Administration.
10 Isotope with mass number 60 is gamma ray source.
14 Device that can be negatively charged and then discharged by ionizing radiation.
20 Studied the penetrating power of α rays
21 _____ effects of radiation that are passed on to following generations.
22 Describes radioactive decay products.
23 Equal to 37,000 dps.

Down

1 Time period for one half of radioactive matter to undergo decay.
2 Equals the # protons plus # of neutrons.
3 Atoms with same atomic number but different atomic mass.
4 Equal to 37 billion dps.

5 The 238 in $^{238}_{92}$U is the _____ number.
6 Irene and Frederic _____ produced the first artificial radioactive isotope in 1934.
7 Sometimes called the "antielectron."
8 Radioisotopes that can be followed to determine path of element.
11 Radioactive noble gas with high density and half-life of 3.82 days.
12 Discovered x-rays in 1896.
13 Roentgen equivalent man, measures biological effects of radiation.
15 A sequence of decay steps in the natural conversion of U-238 to Pb-206.
16 _____ number equals number of protons.
17 Radiation effect limited to the exposed population.
18 Positron emission tomography.
19 Discovered gamma rays in 1900.

Bridging the Gap
Concept Report Sheet
Your Annual Dose and the U.S. Annual Average

Name _____

You can estimate your annual dose using the information on page 136 of your text. This table is derived from that discussion. These tables are adapted from A.R. Hinrichs: Energy, pp 335-336. Philadelphia, Saunders College Publishing, 1992.

1.   Estimate of your annual dose

	Sources of Radiation	Annual dose in mrem
Where You Live	Location: Cosmic radiation at sea level For your elevation (in feet) add this number of mrem Elevation mrem Elevation mrem Elevation mrem 1000 2 4000 15 7000 40 2000 5 5000 21 8000 53 3000 9 6000 29 9000 70 Ground: U.S. average House construction: For stone, concrete, or masonry building add 7	26 ____ ____ 26 ____ ____
What You Eat, Drink, and Breathe	Food, water air: U.S. average Weapons test fallout	24 ____ 4 ____
How You Live	X ray and radiopharmaceutical diagnosis: Number of chest x rays _____ x 10 Number of lower gastrointestinal tract X rays _____ x 500 Number of radiopharmaceutical examinations _____ x 300 (Average dose to total U.S. population = 92 mrem)	____ ____ ____
	Jet plane travel: For each 2500 miles add 1 mrem	____
	TV viewing: Number of hours per day ____ x 0.15	____
How Close You Live to a Nuclear Plant	At site boundary: average number of hours per day _____ x 0.2 One mile away: average number of hours per day _____ x 0.02 Five miles away: average number of hours per day _____ x 0.002 Note: Maximum allowable dose determined by "as low as achievable"(ALARA) criteria established by the U.S. Nuclear Regulatory Commission. Experience shows that your actual dose is substantially less than these limits.	____ ____ ____
	Your total annual dose in mrem	____

2. Comparing your estimated dose with the U.S. Annual Average
    The average annual dose in the U.S. is 180 to 200 mrem. How does your annual dose compare with this average? Is your dose high or low?

3. Principle source of exposure
    What source is the largest contributor to your annual dose?

4. How exposure changes for a frequent flyer
    How many additional mrem would a frequent flyer accumulate if the person took the following trips?   1 mrem per 2500 miles

Trip	Number of round trips	One way mileage
Chicago to San Francisco	2	2108
Chicago to Miami	2	1338
Chicago to Los Angeles	4	1990
Chicago to Seattle	2	2043
Chicago to New York	4	794
Chicago to New Orleans	2	925
Chicago to Denver	1	1037
Chicago to Atlanta	2	695
Chicago to Phoenix	3	1776

What percentage would this be of the average annual dose of 180 mrem?

Bridging the Gap   II   Name _____
Internet access to the
Environmental Protection Agency, EPA
A Citizens Guide to Radon

The EPA home page
The Environmental Protection Agency is readily accessible using the internet. The EPA home page has the following universal resource locator (URL) address,

http://www.epa.gov/

The content of any displayed web page can be down loaded to disk as text or it can be printed. The underlined subject headings on EPA home page are interactive and lead to additional sites. The interactive topics can be selected by clicking on the underlined text. Each one of these active lines leads to another page with a different URL address. The Citizen Information page can be reached if you click on the line."•EPA Citizen Information ". This page can be reached directly by using the following URL.

http://www.epa.gov/epahome/Citizen.html

The "EPA Citizen Information" page
On June 5, 1996 the EPA Citizen Information page looked like the illustration shown on the following page. Remember these pages are constantly updated. The appearance and content of a site will change when the site is updated.

Again this web page has underlined interactive subject headings. Your responsibility is to open the topic " •Publications on Radon". This page has the following direct URL.

http://www.epa.gov/docs/RadonPubs/

Once you have reached this site, scroll down to find and open the EPA page "•A Citizens Guide to Radon". This site contains information about deaths caused by radon, EPA standards for hazardous radon activity levels, plus suggestions on how to reduce radon levels. You are supposed to find the answers to these questions on the Concept Report Sheet.

You should also record your opinion of the EPA web page. Please explore the EPA pages and record the name of a site that was interesting to you. Record the URL for the page you liked best and tell what you liked about it.

EPA Citizen Information

- •National Drinking Water Week is May 5-11!
- •Water Drops: A Collection Just for Kids
- •Safe Drinking Water Hotline
- •EPA Wetlands Information Hotline
- •WATERSHED '96: Moving Ahead Together
- •You and Clean Water
- •Dos and Don'ts Around the Home
- •Consumer Handbook for Reducing Solid Waste
- •Environment Finance Program Publications
    - • Alternate Financing Mechanisms Report
    - • Funding State Environmental Program Administratio: The use of Fee Based Programs
- •UV Index Document
- •Putting Customers First: EPA's Customer Service Plan
- •Guide to environmental Issues
- •Energy Star Computers Product Listing
- •Volunteer Monitoring Program
- •Protect Your Family From Lead in Your Home
- •Terms of Environment
- •Publications on Radon
- •Superfund Program

[EPA Home Page | Comments | Search | Index]

You should open other categories that interest you and record your opinion of the EPA material. Record the URL for the site you liked best and give your reasons for the choice.

Bridging the Gap II
Concept Report Sheet

Name _____

Internet access to the Environmental Protection Agency, EPA
A Citizens Guide to Radon

1. Give the universal resource locator (URL) for the EPA site " •<u>A Citizen's Guide to Radon</u>"

   http://_____

   What does the EPA say about the number of deaths per year that are the result of radon exposure?

2. The EPA gives an estimate of the ratio of radon contaminated homes to clean homes. What is the EPA's estimate?

3. Identify any EPA sites or categories you opened?

4. Give the universal resource locator (URL) for the EPA site you found most interesting.

   http://_____

   What makes it interesting to you? Would you recommend it to someone else? You can send your opinion to the EPA. They want comments.

Solution to Speaking of Chemistry crossword puzzle
Applications of Radioactivity

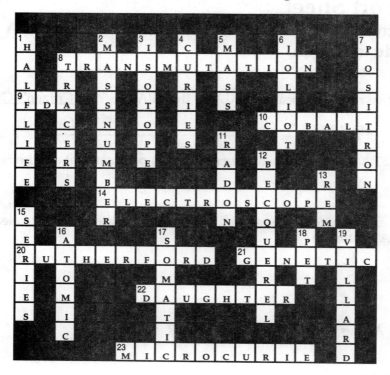

6   Chemical Bonds: The Ultimate Glue

6.1   Ionic Bonds

In this section salts are defined; the octet rule is applied in the formation of Na^+ and Cl^- from the atoms. Ionic bonds and ionic compounds are defined, and the concept of a formula unit is explained. Ionic lattices are describe; ion sizes for cations and anions are listed and compared to the atomic radii of their parent atoms.

Objectives

After studying this section a student should be able to:

      state the definition of a salt
      give definitions for ionic bond and ionic compound
      tell what forces hold an ionic crystal together
      explain why ion sizes differ from atom sizes
      explain why a cation is smaller in diameter than its parent atom and why an anion is larger
          than its parent atom
      tell how ion size relates to parent atom size for a given anion or cation

Key terms

salts	valence electron	metal	nonmetal
octet rule	ionic compound	ionic bond	electrostatic attraction
formula unit	electrostatic forces	ion sizes	electron cloud

6.2   Ionic Compounds

This section relates the octet rule to the Lewis dot symbols for positive and negative ions, shows how ions combine to form neutral compounds, and gives general rules for the formation of ionic compounds and for finding their formulas. It gives the rules for naming binary ionic compounds. Finally, it shows how polyatomic ions combine to form neutral formula units.

Objectives

After studying this section a student should be able to:

      write the Lewis dot symbol for a neutral atom
      state the octet rule
      decide how many electrons an atom will loose or gain to form an octet and predict the
          charge for the ion formed when these electrons are lost or gained
      write the Lewis dot symbol for the anion or the cation formed from a particular atom
      predict the formula for the binary compound formed by a given pair of elements
      give the formula and charge for common ions in the A groups
      give the definition for a binary compound
      explain how anion names are formed from element names for monoatomic (one-atom)
          anions
      state the definition of polyatomic ion
      predict the formula for a combination of a given metal cation and polyatomic anion
      name a binary compound from its formula
      write the formula for a binary compound from its name
      name the common polyatomic anions like those in Table 6.1

Key terms

Lewis dot symbol	noble gas configuration	A group nonmetal
anion	negative ion	"ide" ending
A group metal	cation	Roman numeral and charge
binary compound	ammonium ion	polyatomic ion
carbonate ion	hydrogen carbonate	hydroxide ion
phosphate ion	nitrate ion	nitrite ion
sulfate ion	sulfite ion	acetate ion
hydrogen sulfate ion	cyanide ion	hydrogen phosphate ion
hydrogen sulfite ion	dihydrogen phosphate ion	hypochlorite ion

6.3   Covalent Bonds

This section gives a definition for the covalent bond, describes single covalent bonds, defines unshared (lone pair, nonbonding) electrons, and shows how to predict molecule formulas for nonmetal-nonmetal combinations like $:\!\overset{\displaystyle\cdot}{\underset{\displaystyle\cdot}{N}}\!\cdot$ and $H\cdot$ to make $H\!:\!\overset{\displaystyle\cdot\cdot}{\underset{\displaystyle\cdot\cdot}{N}}\!:\!H$ with an H above. It describes saturated hydrocarbons, gives the formula C_nH_{2n+2} for saturated hydrocarbons, describes multiple covalent bonds (double and triple bonds), summarizes alkenes and alkynes or unsaturated hydrocarbons, and defines bond energy. The section describes how to name binary molecular compounds, and gives definitions for the prefixes used in naming. It shows how to relate single, double and triple bonds to Lewis dot structural formulas for molecules.

Objectives

After studying this section a student should be able to:

state the definition for a covalent bond

draw the Lewis symbol for a molecule if given the formula such as H_2O

describe the difference between a nonbonding pair of electrons and a bonding pair

give definitions for hydrocarbons, alkanes, and saturated hydrocarbons

describe single, double, and triple covalent bonds

give a definition for bond energy and explain what the values represent

tell how alkanes differ from alkenes and alkynes

describe what unsaturated means

identify the types of bonds in a molecule if given the structural formula

match prefixes and number of atoms i.e.. mono- one; di- two; tri- three

name representative binary molecular compounds if given the formula

Key terms

Lewis structure	covalent bond	single covalent bond
nonbonding pair	bonding pair	hydrocarbon
alkane	saturated hydrocarbon	C_nH_{2n+2}
multiple covalent bond	double bond	triple bond
nitrogen fixation	bond length	bond energy
alkene	unsaturated hydrocarbon	C_nH_{2n}
binary molecular compounds	prefixes	mono-
di-	tri-	tetra-
penta-	hexa-	hepta-
octa-	nona-	deca-
common names	ammonia	

6.4   Guidelines for Drawing Lewis Structures

This section gives a four-step procedure for constructing Lewis structures for molecules and ions using the formula and the periodic table:
1) identify the center atom and the terminal atoms,
2) count the total number of valence electrons for all atoms,
3) draw skeletal structure and place a pair of electrons between each two bonded atoms,
4) place remaining valence electron pairs around atoms to fulfill octet rule for each (except H); move electron pairs to form multiple bonds if needed to satisfy the octet rule.

Objectives
After studying this section a student should be able to:

give the count for the valence electrons in a molecule like CH_3OH

give the number of valence electrons in a polyatomic ion such as CO_3^{2-} (it has 24)

draw the Lewis dot structure for a molecule or ion if given the formula

identify the central atom in a formula

recognize that large molecules like the one shown consist of many "central atoms"

has 4 "central" atoms

Key terms

arrangement of atoms	terminal-atom	halogen
binary compound	central atom	lowest subscript
valence electrons	skeletal structure	nonbonding pairs
bonding pairs	bonded atoms	ion charge

6.5   Exceptions to the Octet Rule

This section gives examples of compounds that do not follow the octet rule. Three classes of compounds are illustrated: compounds with fewer than 8 valence electrons, such as BH_3, BeH_2; compounds with expanded valence where the central atom is in period 3 or higher, such as PF_5, SF_6; and compounds with an odd number of valence electrons, such as NO.

Objectives
After studying this section a student should be able to:

name the three types of compounds that do not follow the octet rule

give a reason why each of the three types deviates from the octet rule

tell if a compound is an exception to the octet rule by checking its Lewis structure

give definitions for expanded valence, odd-electron compound

Key terms

expanded valence   odd-electron compounds   fewer than eight electrons

6.6   Predicting Shapes of Molecules and Polyatomic Ions

This section explains the VSEPR model for predicting shapes of molecules and ions, and reviews definitions of bonding and nonbonding pairs of electrons. It introduces bond angle measurements and describes the linear, bent, triangular planar, triangular pyramidal and tetrahedral shapes. The section also shows how the AXE formula is used to predict a molecule or ion's shape from its Lewis structure.

Objectives

After studying this section a student should be able to:

give the definition for "VSEPR"

state the logic behind VSEPR

predict the shape of a molecule or ion, i.e., its molecular geometry, by drawing its Lewis structure and applying the VSEPR theory

state the difference between the terms "electron-pair geometry" and "molecular geometry"

use the electron-pair geometry to predict bond angles

identify the AXE class for a central atom in a formula after drawing its Lewis structure

use the AXE class to predict the shape of the molecule or ion

Key terms and equations

valence-shell electron-pair repulsion theory

linear geometry

tetrahedral geometry

molecular geometry

VSEPR

triangular planar geometry

electron pair geometry

bent geometry

6.7 Polar and Nonpolar Bonding

Linus Pauling's Electronegativity scale is described and electronegativity values for the A Group elements are given. Bond polarity is explained for nonpolar and for polar covalent bonds. Molecular shapes are related to bond polarity to explain the polarity of molecules. The effect of symmetry on the polarity of molecules is described. The convention of using $\mapsto$ to indicate bond polarity is introduced.

Objectives

After studying this section a student should be able to:

state the definition for electronegativity

tell what atom is the most electronegative one in a compound given the formula

describe how electronegativity changes for elements in the same group

describe the trend for electronegativity across the periodic table

give the basis for deciding on the polarity of a bond

show the polarity of a bond for a given pair of atoms using the $\mapsto$ symbolism

describe the difference between a polar and a nonpolar bond

tell if a molecule is polar or nonpolar using electronegativities and molecule shape

tell what "equal sharing" means in covalent bonding

Key terms and equations

electronegativity

equal sharing

polar molecule

symmetrical

nonpolar covalent bond

unequal sharing

electron shift

polar covalent bond

nonpolar molecule

partial charges

6.8 Properties of Molecular and Ionic Compounds Compared

This section explains the reasoning behind the axiom that "like dissolves like". It describes the physical properties of polar and nonpolar compounds, summarizes the properties of ionic and molecular compounds, and introduces the concept of electrolytes and nonelectrolytes in terms of conductivity.

Objectives

After studying this section a student should be able to:

tell what the "like dissolves like" rule means

outline the properties of ionic and molecular compounds in terms of reactivity, hardness, melting point range, boiling point range, solubility in water, and conductivity of the substance when molten (melted)

Key terms and equations

like dissolves like	brittle	crystalline solid
molecular compound	ionic compound	conductivity
current	electrolyte	nonelectrolyte
low boiling point	high boiling point	

Additional readings

Amoore, John E.., et al. "The Stereochemical Theory Of Odor" <u>Scientific American</u> Feb. 1964: 42.

Breslow, Ronald. "The Nature Of Aromatic Molecules" <u>Scientific American</u> Aug 1972: 32.

Curl, Robert F. and Richard E. Smalley. "Fullerenes" <u>Scientific American</u> Oct. 1991: p 54.

Derjaguin, Boris V. "The Force Between Molecules" <u>Scientific American</u> July, 1960: p 47.

Kalmus, Hans. "The Chemical Senses" <u>Scientific American</u> April 1958: p 97.

Lambert, Joseph B. "The Shapes Of Organic Molecules" <u>Scientific American</u> Jan 1970: p 58.

Turner, Barry E. "Interstellar Molecules" <u>Scientific American</u> Mar 1973: p 50.

Zewail, Ahmed H. "The Birth Of Molecules" <u>Scientific American</u> Dec. 1990: p 76.

Answers to Odd Numbered Questions for Review and Thought

1.
   a. A cation is an ion with a positive charge, like Ca^{2+}.
   b. An anion is an ion with a negative charge, like Cl^{1-} or CO_3^{2-}.
   c. Atoms react to acquire an electron configuration with 8 electrons in the outermost level.
   d. The formula unit is the simplest element ratio for an ionic compound like NaCl instead of Na_2Cl_2.

3.
   a. A shared pair is a pair of electrons shared between two atoms like in H:Cl
   b. Four electrons (2 pairs) shared by two atoms as between carbons in H_2C :: CH_2
   c. Six electrons (3 pairs) shared between two atoms as in HC ::: CH
   d. A pair of valence electrons on an atom that are not shared with another atom.
   e. A single bond is one pair of electrons shared between two atoms as in H:H
   f. A multiple bond is either a double or a triple bond.

5. a. A nonpolar bond exists between two atoms that share electrons equally.
    b. A polar bond exists between two atoms that do not attract shared electrons equally.
    c. Electronegativity is a measure of an atom's attraction for shared electrons.

7. a. A compound consisting only of carbon and hydrogen like ethane, CH_3CH_3.
    b. A compound consisting only of carbon and hydrogen with only single bonds between carbon atoms
    c. A compound consisting only of carbon and hydrogen with one or more multiple bonds between carbon atoms
    d. A molecule with formula C_{60} that looks like a soccer ball.
    e. Alkenes are hydrocarbons with one or more carbon-carbon double bonds.

9. a. Bromine is in Group VIIA and has seven valence electrons so it gains one to form Br^{1-}.
    b. Aluminum is in Group IIIA and has three valence electrons so it loses 3 to form Al^{3+}.
    c. Sodium is in Group IA and has one valence electrons so it loses one to form Na^{1+}.
    d. Barium is in Group IIA and has two valence electrons so it loses two to form Ba^{2+}.
    e. Calcium is in Group IIA and has two valence electrons so it loses two to form Ca^{2+}.
    f. Ga is in Group IIIA like aluminum so it loses its three valence electrons to form Ga^{3+}.
    g. Iodine is in Group VIIA like Br so it has seven valence electrons and it forms I^{1-}.
    h. Sulfur is in Group VIA so it has six valence electrons and gains two electrons to form S^{2-}.
    i. Group IA atoms lose one valence electron to form a +1 ion, see 9c above.
    j. Group VIIA atoms have seven valence electrons and gain one to form a 1- ion, see 9g.

11. The 16 protons indicate the element is sulfur. The charge is -2 because there are 2 more electrons than protons. The symbol would be S^{2-}.

13. a. Neutral atom. A cation has more protons than electrons so the electron cloud is pulled in closer to the nucleus. The electrostatic attraction for the electrons by the nucleus is greater in the cation than in the neutral atom.
    b. Anion. Additional electrons are repelled by other electrons in the electron cloud and there are more electrons than protons so each electron is held less tightly by the nucleus. The repulsions between electrons in an anion are greater than in the neutral atom.

15. Electrostatic attractions between positive and negative ions in an ionic lattice holds the solid together. These attractions are not limited to the nearest ions. The example is an NaCl lattice.

17.

Formula	MP	Formula	MP	Formula	MP
NaCl	801 °C	$MgCl_2$	708 °C	$AlCl_3$	190 °C
Na_2S	1180 °C	MgS	decomposes	Al_2S_3	1100 °C
Na_3P	decomposes	Mg_3P_2	not listed	AlP	not listed

All are solids at room temperature. For a given cation the melting points increase with the charge on the anion.

19. These are generalizations and there will be some exceptions.

Group IA	II A	III A	IV A	V A	VI A	VII A,	VIII A
H 1 bond	no	3 bonds	4 bonds	3 bonds	2 bonds	1 bond	0 bonds
(Na-Fr do not form co-valent bonds)	covalent bonds						

21. The charge on Tb is determined by the -3 charge on the phosphate ion, PO_4^{3-} in $Tb_3(PO_4)_4$, because the total charge must add to zero.
The positive charge on terbium = X. The charge on phosphate is -3. The total charge on all the ions in the phosphate compound must be zero, 0.
This means $3X + 4(-3) = 0$; $3X - 12 = 0$; $3X = 12$; $X = 4$.
The terbium ion is Tb^{4+}. The expected formula is $Tb(SO_4)_2$. The combination of sulfate and terbium ion should be ionic because the sulfate ion is stable and will not break apart if combined with a positive ion like Tb^{4+}.

23. a. $Al_2(SO_4)_3$ aluminum sulfate   b. $Ca(H_2PO_4)_2$ calcium dihydrogen phosphate
    c. K_3PO_4 potassium phosphate   d. NH_4NO_3 ammonium nitrate
    e. Na_2CO_3 sodium carbonate   f. $CaHPO_4$ calcium hydrogen phosphate

25. a. An atom with an electron arrangement of 2-8-1 has one valence electron and will lose 1 electron to form a +1 cation.
    b. An atom with an electron arrangement of 2-7 is in Group VIIA with 7 valence electrons and will gain 1 electron to form a -1 anion.
    c. An atom with an electron arrangement of 2-4 is in Group IVA with 4 valence electrons and will gain 4 electrons.
    d. An atom with an electron arrangement of 2-8-8-2 will be in Group IIA and will lose 2 electrons to form a +2 cation.

27. a. Sodium is in Group IA, NaH   b. Magnesium is in Group IIA, MgH_2
    c. Gallium is in Group IIIA, GaH_3 d. Germanium is in Group IVA, GeH_4
    e. Arsenic is in Group VA, AsH_3 f. Chlorine is in Group VIIA, HCl

29. a. nitrogen monoxide   b. sulfur trioxide
    c. dinitrogen oxide   d. nitrogen dioxide

31. a. Hydrogen peroxide H_2O_2, has 14 valence electrons.

Valence electrons from hydrogen	Valence electrons from oxygen	Total valence electrons
2 x 1 +	2 x 6 =	14

b. Ammonia, NH_3, has 8 valence electrons.

Valence electrons from hydrogen	Valence electrons from nitrogen	Total valence electrons
3 x 1 +	1 x 5 =	8

c. Hydrogen sulfide, H_2S, has 8 valence electrons.

Valence electrons from hydrogen	Valence electrons from sulfur	Total valence electrons
2 x 1 +	1 x 6 =	8

33. The Lewis structure first requires counting valence electrons including any adjustments for ionic charges. Add electrons for negative charge and delete electrons for positive charge. Then central atoms must be identified. The terminal atoms are arranged around the central atoms. Electrons are placed to form single bonds between atoms to satisfy octets. If single bonds do not meet the octet requirements multiple bonds should be tried.

a. N_2H_4

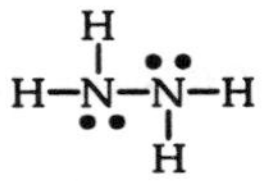

b. CH_3OH

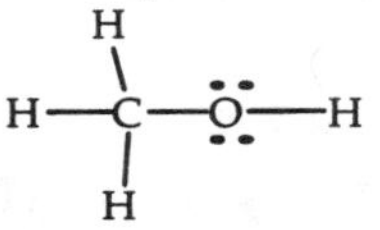

c. BCl_3

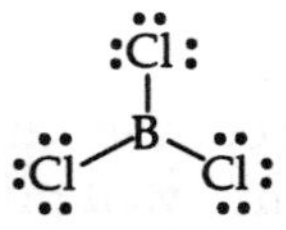

d. PH_3

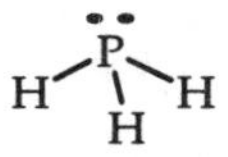

e. $ClO_3{}^{1-}$

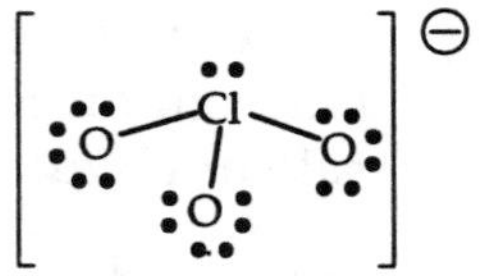

f. $SO_3{}^{2-}$

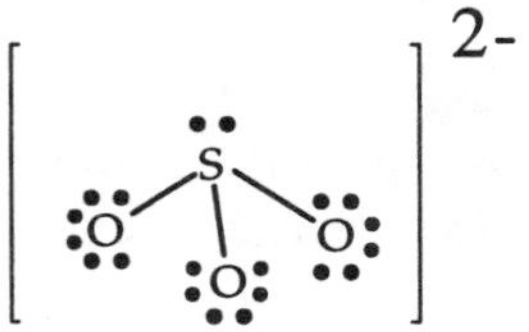

g. $NH_4{}^{1+}$

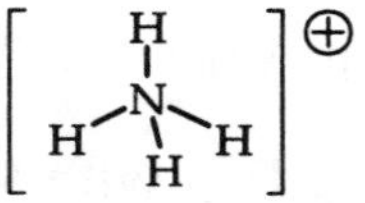

h.

35. The electron-pair geometry depends on the positions of the electron pairs around the center atom in the molecule. The molecular geometry refers to the shape of the molecule; shapes are described in terms of locations of atoms (nuclei) in the molecule.
Water: electron-pair geometry is tetrahedral and molecular shape is angular or bent.

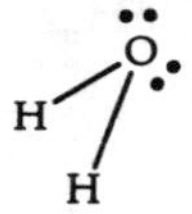

37.

Lewis formula	electron-pair geometry	molecular geometry
a. 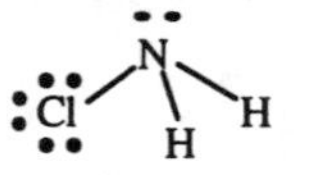	tetrahedral	trigonal pyramid
b. 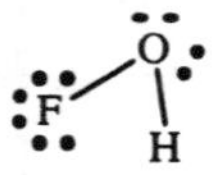	tetrahedral	angular
c. 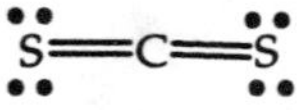	linear	linear
d. 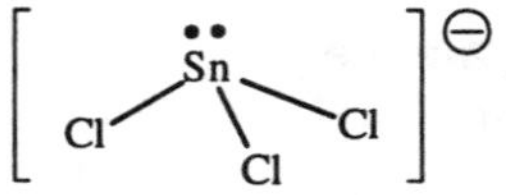	tetrahedral	trigonal pyramid

e.   tetrahedral   tetrahedral

f.   tetrahedral   tetrahedral

39.   H——C≡N: linear

41.   a.   The electronegativity differences determine the polarity of the bonds. **The**
     electronegativity for carbon is 2.5 for hydrogen it is 2.1 and it is 3.0 for chlorine. **The**
     differences are C-Cl   3.0 - 2.5 = 0.5 and for C-H 2.5 - 2.1 = 0.4 The C-Cl **bond is more**
     polar because it has a bigger electronegativity difference.
    b.   The C-F bond with an electronegativity difference of 4.0 - 2.5 = 1.5 **is more polar than**
     the C-Cl bond with a difference of 0.5.

43. $BeCl_2$ is linear so partial charges on   in the symmetric molecule. **It is a**
    the chlorine atoms counteract each   **nonpolar molecule even though the**
    other. **The center of the two positive**   bonds are polar.
    partial charges falls on the center of
    the Be nucleus. The center of charge
    for the two negative partial charges
    falls on the center of the Be nucleus
    also. There is no separation of charge

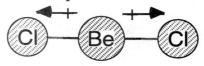

45.   a. HF

47.   The electron-pair geometry is basically tetrahedral but the lone pair on **the N distorts the**
    HNH angle from 109.5° to 106.5°.

49.   a.   **There are 24 valence electrons in acetone**

valence electrons from hydrogens		valence electrons from carbons		valence electrons from oxygen		total electrons
6 x 1	+	3 x 4	+	1 x 6	=	24

    b.   **There are 8 single covalent bonds: six C-H single bonds and two C-C single bonds**
    c.   **There is one double bond between C and O**
    d.   **There are 10 bonding electron pairs.** There are 8 bonding pairs from **the single bonds**
      and 2 bonding pairs from the C::O **double bond.**
    e.   There are two nonbonded pairs of electrons on the oxygen atom.

    Lewis symbol structure   Ball and stick model   Space filling model

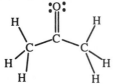 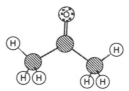 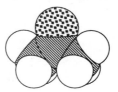

Solutions for Selected Problems

1. The cation diameter equals two times the cation radius. diameter = 2 x radius;

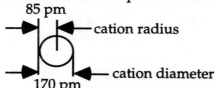

85 pm — cation radius

170 pm — cation diameter

If the cation radius is 85 picometers then the diameter = 2 x 85 pm = 170 pm, Similarly the anion diameter = 2 x radius; anion diameter = 2 x 145 pm = 290 pm

2. The packing of ions in a solid can be simulated in one dimension by a linear arrangement of coins representing ions in the ionic crystal. Typically cations (positive ions) have a smaller diameter than anions. The quarter is the 290 pm diameter anion and the penny is the 170 pm cation. The total length of all the ions or coins = 1670 pm.

total
length = quarter + penny + quarter + penny + quarter + penny + quarter
1670 pm = 290 pm + 170 pm + 290 pm + 170 pm +290 pm + 170 pm +290 pm

Speaking of Chemistry

Name _____

Chemical Bonds

Across

1 Prefix meaning one.
3 Prefix meaning five.
5 Valence shell electron pair repulsion theory.
7 Particle with positive charge.
10 _____ dissolves like.
11 Shape for CO_2, carbon dioxide.
12 Negatively charged ion.
14 Measure of an atoms attraction for electrons in a bond.
18 _____ valence exists on S in SF_6.
21 Number of bonds formed by carbon.
24 Hydrocarbons with at least one double bond.
25 A count of electrons like 1,3 , etc.
27 The number of bonds in N_2.
28 Electrons in a triple bond.

Down

2 Positive charge on Group IA ions.
4 Number of electrons in single bond.
6 Name for the S^{2-} ion.
8 Compound with formula NH_3.
9 Ending on single atom anion names.
13 Prefix meaning six.
15 Name for Cl^{1-} ion.
16 A set of eight electrons.
17 Bond _____ for C-C is 83 kcal/ mol.
19 Hydrocarbons with only carbon carbon single bonds.
20 Prefix meaning two.
22 Shape of water molecule.
23 Prefix meaning four.
26 The prefix meaning three.

Bridging the Gap
Odors, Perfumes, and Molecular Shapes

There is evidence that the sense of smell is based on the geometry of molecules. The nose can easily tell the difference between a skunk and a rose. Seven primary odors are distinguished by a specifically shaped receptor at the olfactory nerve ending. Each primary odor is associated with a receptor site. The seven primary odors are floral, musky, camphoraceous, pepperminty, ethereal, pungent and putrid. Complex odors result when more than one type of receptor site is stimulated. The shapes for the receptor sites are illustrated below. Example molecules are shown that fit into the respective receptors. The result is that unrelated compounds can trigger the same odor because the geometric shapes for the molecules are in the same class. Five receptors are sensitive to shape and two receptors are sensitive to electrostatic forces. One is attracted to molecules that have an electron rich region and the other is attracted to molecules that are have a localized positive charge.

The purpose of this exercise is to give you some practice in working with molecular shapes. You are supposed to pretend that you are trying to formulate a perfume. It is your job to select compounds that will be good candidates for use in the perfume from a list of compounds and their molecular space filling models. You are supposed to identify the odor expected for each compound given as potential fragrances and tell which ones you would use for the perfume.

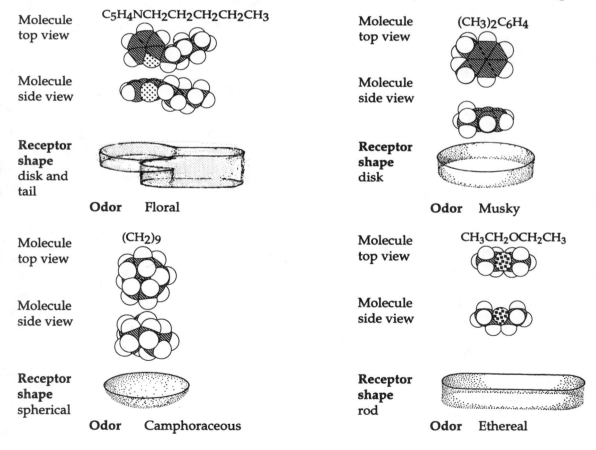

Molecule top view $C_5H_4NCH_2CH_2CH_2CH_2CH_3$

Molecule side view

Receptor shape disk and tail

Odor Floral

Molecule top view $(CH_3)_2C_6H_4$

Molecule side view

Receptor shape disk

Odor Musky

Molecule top view $(CH_2)_9$

Molecule side view

Receptor shape spherical

Odor Camphoraceous

Molecule top view $CH_3CH_2OCH_2CH_3$

Molecule side view

Receptor shape rod

Odor Ethereal

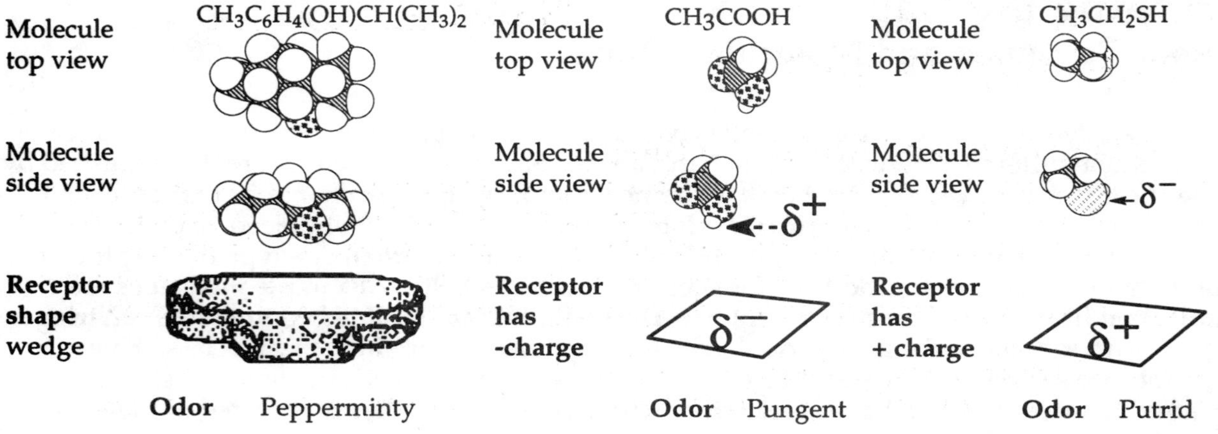

Molecule top view	$CH_3C_6H_4(OH)CH(CH_3)_2$	**Molecule top view**	CH_3COOH	**Molecule top view**	CH_3CH_2SH

Molecule top view $CH_3C_6H_4(OH)CH(CH_3)_2$

Molecule side view

Receptor shape wedge

Odor Pepperminty

Molecule top view CH_3COOH

Molecule side view ←--δ^+

Receptor has -charge δ^-

Odor Pungent

Molecule top view CH_3CH_2SH

Molecule side view ←δ^-

Receptor has + charge δ^+

Odor Putrid

Bridging the Gap
Concept Report Sheet

Name _____

Odors, Perfumes, and Molecular Shapes

The following table of compounds are available to you for the formulation of a new perfume. Your responsibility to identify the odors each molecule will stimulate. You are to point out the structural features that made the molecule fit into the respective categories. Remember a molecule type may fit more than one receptor class. You are supposed to make your selection of compounds for your product and tell what primary odors you are using in your product.

1. Matching Molecular Shapes, Receptor Types, and Odor

	$CH_3CH_2CH_2CH_2SH$	$CH_3OCH_2CH_3$	$C_6H_5CH_2CH_2CH_2CH_3$	$C_{13}H_{24}O$
Molecule top view				
Molecule side view				
Receptor site type (more than one may be a appropriate)				
Odor				

2. Choices for Making a Perfume
    Which of the available compounds would be appropriate for use in a perfume? Justify your choices.

Solution to Speaking of Chemistry crossword puzzle
Chemical Bonds

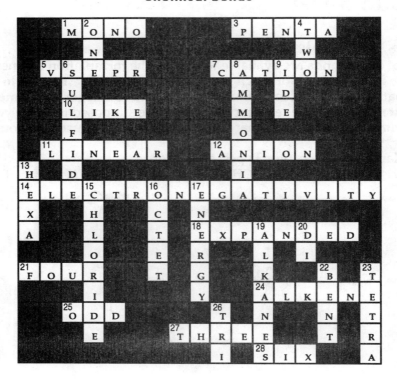

7 States of Matter and Solutions

7.1 The Kinetic Molecular Theory: How and Why Does Matter Change State?

The Kinetic Molecular Theory postulate that molecules in solids, liquids, and gases are in constant motion is described. The distances traveled by particles in solids, liquids, and gases are compared. The relationship between temperature and movement is explained.

Objectives
After studying this section a student should be able to:
 tell how the freedom of movement of particles differs for solids, liquids, and gases
 describe how increasing temperature influences particle velocity
 use the KMT to explain why gases expand to fill a container
 use the KMT to explain why liquids are free to flow

Key terms

states of matter	kinetic molecular theory	solid
liquid	gas	regular arrangement
fluid	flow	"particles" as used in KMT

7.2 Gases and How We Use Them

The Kinetic Molecular Theory model for gas pressure and gas compressibility is explained. Boyle's Law is described in terms of KMT. The expansion of gases on heating, Charles' Law, is interpreted in terms of KMT. The Kelvin temperature scale and absolute zero are introduced. Dalton's equation for pressure of a gas mixture is explained. Gas diffusion and liquid volatility are described. The origin of atmospheric pressure and the measurement of barometric pressure are illustrated.

Objectives
After studying this section a student should be able to:
 describe how frequency of gas particle collisions relates to temperature
 summarize Charles' Law

 explain why a real gas cannot have zero volume at $0^\circ K$
 explain why, at constant pressure, gas volume decreases with a decrease in temperature
 state what absolute zero equals on the Celsius scale
 tell how the pressure of each gas in a mixture contributes to the total pressure
 explain why gases are miscible
 state the definition for the term "volatile"
 tell what gaseous diffusion means

Key terms

plasmas	compressed air	SCUBA
pressure	compressibility	Boyle's law
volume	temperature	Charles' law
expansion	Atmospheric pressure	barometer
directly proportional	inversely proportional	absolute zero
Kelvin temperature scale	kelvin or kelvin degree	0 kelvin
miscibility	barometric pressure	Celsius scale
volatile	gaseous diffusion	homogeneous

7.3 Intermolecular Forces

This section gives definitions for intermolecular forces. The attractions between polar molecules are described. The origins of attractive forces between nonpolar molecules are discussed. Requirements for hydrogen bonding between polar molecules like HF and H_2O are described in detail.

Objectives
After studying this section a student should be able to:

give a definition for intermolecular forces

explain what condensation is and how it relates to intermolecular forces

give a rough idea of the strength of intermolecular forces compared to covalent bonds

describe how polar molecules like SO_2 interact using words and sketches

describe the origin of intermolecular forces between nonpolar molecules

tell how molecular size and valence electrons affect forces between nonpolar molecules

give a definition for hydrogen bonding and tell how it affects the properties of H_2O

state the requirements for hydrogen bonding

Key terms

intermolecular forces	attractive forces	condensation
polar molecule	partially negative region	partially positive region
nonpolar molecule	induced attractive forces	hydrogen bonding
highly electronegative atom	electronegativity	tetrahedral cluster

7.4 Water and Other Pure Liquids

This section describes the physical properties of water and relates them to the unusually strong intermolecular (hydrogen bonding) attractions between water molecules. The following properties are reviewed: normal boiling point, density, heat capacity, heat of vaporization, surface tension, and ability to act as a solvent.

Objectives
After studying this section a student should be able to:

explain why ice (solid water) has a lower density than liquid water

give definitions for heat capacity, heat of vaporization and surface tension

give a definition for vapor pressure

tell why the normal boiling point for a liquid indicates volatility

examine a graph of vapor pressure versus temperature for a series of liquids and identify the most volatile

describe how vapor pressure of a liquid depends on temperature

define normal boiling point and boiling point and tell how these differ

Key terms

pure	liquid water	ice
water vapor	density	heat capacity
hydrogen bonding	heat of vaporization	surface tension
solvent	the universal solvent	slightly compressible
vapor pressure	boiling point	normal boiling point
standard pressure	temperature-vapor pressure curves	

7.5   Solutions

This section gives definitions for terms that describe solutions and the dissolving process. The effects of structure on solubility are explained. The effects of temperature on gas solubility are described. The properties and applications of supercritical fluids are discussed.

Objectives
After studying this section a student should be able to:

give definitions for solvent, solute, solubility, solution

state definitions for saturated and unsaturated solutions

describe how solubility of alcohols in water depends on the number of carbons in the alcohol molecules

explain why small alcohol molecules are more water soluble than large ones

describe how gas solubility in water depends on temperature

tell how gas pressure influences gas solubility and give an example

give definitions for critical temperature, critical pressure and supercritical fluid

describe a use for supercritical fluids

Key terms

solution	homogeneous mixture	solvent
solute	solubility	miscibility
saturated	unsaturated	insoluble
dissolve	gas solubility	gas solubility & temperature
supercritical fluids	critical temperature	gas solubility & pressure
critical pressure		

7.6   Solids

This section includes a comparison of the freedom of movement of particles in solids, liquids, and gases. The hardness of talc and of diamond are described and explained. The regular arrangement of particles in solids is outlined. Melting, crystallization, and sublimation are described. The sublimation of CO_2 and the principles behind frost-free refrigerators are discussed. Melting points and intermolecular forces are summarized for different types of solids.

Objectives
After studying this section a student should be able to:

describe the structure for diamond and explain why diamond is very hard

describe the bonding and structure for talc and explain why it is soft

define melting point

describe crystallization

give range of melting points for molecular solids containing nonpolar molecules like H_2

give range of melting points for molecular solids containing polar molecules like HCl

give range of melting points for ionic solids like NaI, NaCl

describe how sublimation of H_2O is used

Key terms

random motion	vibration	rotation
orderly array	crystals	hardness
regular arrangement	molecular motion	melting point
solidification	crystallization	sublimation
sublime		

7.7   Metals, Semiconductors, and Superconductors

This section describes the major characteristics of metals, semimetals (semiconductors) and superconductors. Models that explain these properties are introduced. Applications for semiconductors and superconductors are discussed.

Objectives

After studying this section a student should be able to:

describe and explain electrical conductivity, thermal conductivity, ductility, and luster of metals

tell what the "electron sea" is in metals

give an example of a semiconductor and tell how they differ from metals

tell what doping of nonmetal crystals does to the crystal structure and conductivity

give definitions for p-type and n-type semiconductors

describe the combination of semiconductors used to assemble a transistor

define a superconductor

define alloys

tell how MRI machines and superconductors are related

Key terms

metals	electrical conductivity	thermal conductivity
ductility	malleability	luster
insolubility	lattices	sea of mobile electrons
valence electrons	semiconductors	electron gate
doping	dopant	positive holes
extra electrons	p-type semiconductor	n-type semiconductor
transistor	superconductor	superconducting transition temperature
alloys	MRI	magnetic resonance imaging

7.8   Gas Law Arithmetic

This section shows how to use Boyle's Law, Charles' Law, and the combined gas laws to predict gas properties such as volume, temperature, and pressure.

Objectives

After studying this section a student should be able to:

use Boyle's Law to solve for either an unknown pressure or volume

use Charles' Law to solve for either an unknown volume or temperature

use the combination of Boyle's and Charles' Laws to solve for unknown volume or temperature or pressure

convert Celsius temperatures to Kelvin and the opposite, Kelvin to Celsius

Key terms

Boyle's Law	volume of gas	varies inversely
applied pressure	$P \times V = k$	varies directly
Charles' Law	absolute temperature	$V = kT$
Kelvin scale		$P_1 \times V_1 = P_2 \times V_2$

$$\frac{V_1}{T_1} = \frac{V_2}{T_2} \qquad \frac{P_1 V_1}{T_1} = \frac{P_2 V_2}{T_2} \qquad P\alpha\frac{1}{V}$$

Additional readings

Amato, Ivan. "New Superconductors: a Slow Dawn" <u>Science</u> Jan. 15, 1993: 306

Coy, Peter. "Man-made Diamonds Learn a New Trick from Mother Nature" <u>Business Week</u> June 7, 1993: 103.

Colson, Steven D. and Thom H. Dunning Jr. "The Structure of Nature's Solvent: Water" <u>Science</u> July 1, 1994 : 43.

Hansen, Lars. "Weird Fluids, Big Business: Supercriticals May Be Reaching Critical Mass" <u>Business Week</u> Sept. 11, 1995: 100

Keyes, Robert W. "The Future of the Transistor" <u>Scientific American</u> June 1993: 70.

Maddox, John. "Paper Crystals of Molecular Hydrogen and Ice" <u>Nature</u> August 5, 1993: 483.

Pool, Robert. "Atom Smith (Dick Siegel Constructs Materials One Molecule at a Time)" <u>Discover</u> Dec. 1995: 54.

Posterino, P. et al. "The Interatomic Structure of Water at Supercritical Temperatures" <u>Nature</u> Dec. 16, 1993: 668.

Taylor, Larry. "SFE Provides Quick Release of Stubborn Compounds" <u>R & D</u> Feb.1992: 78.

Answers to Odd Numbered Questions for Review and Thought

1.   a.   The gaseous state has no definite shape or volume; a gas takes the shape and volume of its container. Gas molecules have high kinetic energy compared to the attractive forces between them and are therefore, free to move away from each other until the gas has filled the container. The distances between gas molecules are great, almost 1000 times the size of the molecules; this makes gases compressible.
     b.   A liquid has definite volume and assumes the shape of the container. Molecules in a liquid are in contact with each other and attractions exist between them, keeping the liquid volume the same. The liquid molecules are free to move past each other, allowing the liquid to flow and assume the shape of its container. Because molecules are in contact already, there is very little room for them to move any closer together, and liquids are relatively incompressible.
     c.   The solid state has definite shape and definite volume. The molecules or ions making up the solid are in direct contact and are occupying fixed positions in the solid arrangement, so shape and volume stay the same for a solid. The solid is incompressible due to direct contact between the particles.

3   A gas has the greatest average distance between particles. Gas compressibility is evidence for this.

5.   Particles in a liquid are in constant motion; this is shown when one liquid diffuses in another (a drop of ink or food coloring will slowly spread throughout a glass of water even though you do not stir the mixture). Particles in a liquid are close together. This makes liquids nearly incompressible.

7. The number of gas molecules in a given volume of gas decreases as you move away from Earth's surface. Additionally, the force of Earth's gravity decreases with altitude. This results in less weight or force per unit area.

9. The pressure exerted by the gas mixture results from the collisions of molecules of both gases with the container walls. When both gases are placed in the same container, molecules of each gas will act independently and will strike the container walls, exerting their pressure just as if the other gas were not present. The pressure of the mixture will be the pressure from gas A <u>plus</u> the pressure from gas B.

11.

The hydrogen bond (shown as a faint line) in water exists between a hydrogen atom in one water molecule and an oxygen atom in a different water molecule. One oxygen is typically hydrogen bonded to two different water molecules.

13. a. Yes. Hydrogen (electronegativity 2.1) is covalently bonded to a very electronegative oxygen atom (electronegativity 3.5). The hydrogen in the O-H group can hydrogen bond to an oxygen in another molecule.
    b. Yes. In ammonia, hydrogen is covalently bonded to a very electronegative nitrogen atom (3.0). The N-H bond is very polar and hydrogens in one NH_3 can hydrogen bond to a nitrogen in a different NH_3 molecule.
    c. No. Sulfur dioxide, SO_2, has no hydrogens so no hydrogen bonding is possible even though it is a bent polar molecule.
    d. No. Carbon dioxide, CO_2, has no hydrogens so it cannot show hydrogen bonding. It is a linear nonpolar molecule.
    e. No. Methane does not show hydrogen bonding even though it has four hydrogens. The electronegativities of hydrogen (2.1) and carbon (2.5) are similar. The C-H bond is not polar enough to produce hydrogen bonding between methane molecules.
    f. Yes. Hydrogen is covalently bonded to fluorine in HF. The electronegativities are very different hydrogen (2.1) and fluorine (4.0). This is the most polar bond formed by a hydrogen atom. The hydrogen in one HF molecule can hydrogen bond to fluorines in neighboring HF molecules.
    g. No. The hydrogens are bonded to the carbon atoms not the oxygen. The electronegativity of carbon (2.5) and hydrogen (2.1) are similar so the bonds are not polar enough to produce hydrogen bonding.

15. Water molecules have strong attractions for one another. The water molecules making up the surface are difficult to separate and the surface seems relatively hard to the "belly flopper".

17. There are great distances between particles in the gas state while particles in the liquid and solid states are in direct contact. The gas molecules are separated by distances equal to about 1000 molecular diameters. This is empty space, a vacuum.

19. The normal boiling point is the temperature at which the vapor pressure of the liquid equals 760 mm Hg, (1 atm).

21. Attractive forces between particles in the surface layer are unbalanced because the particles are attracted by the bulk of the liquid below the surface. Water has a high surface tension because of the strong intermolecular forces between molecules. They are tied to one another by strong dipole-dipole attractions and hydrogen bonding.

23. The boiling point is found by locating the temperature where an imaginary 200 mmHg vapor pressure line crosses the vapor pressure curve, about 70°C.

25. No, the carbon skeleton is too large. The water solubility of alcohols decreases as the number of carbon atoms in the alcohol increases. Methanol, CH_3OH, and ethanol, CH_3CH_2OH, are miscible in water, . When the carbon chain increases to seven carbons, only 0.09 grams of 1-heptanol, $CH_3(CH_2)_5CH_2OH$, dissolves in 100 grams of water at 20°C.

27. The solubility of oxygen in water decreases when water temperature goes up. Fish may not be able to get enough dissolved oxygen from the warmer water.

29. sublimation

31. Water, H_2O, has a density of 1 gram / mL and it floats on carbon tetrachloride, CCl_4, which has a density of 1.59 grams/ mL.

33. Electrical conductivity of metals depends on the easy movement of valence electrons from atom to atom in the metal. The valence electrons are shared by all the atoms of the metal. The valence electrons are not tied down in covalent bonds nor are they part of negative ions.

35. The n-type have excess electrons while p-type have a shortage of electrons. The n-type semiconductors were created by replacing an atom like Si that has four valence electrons with an atom like As, arsenic, that has five valence electrons. The crystal then has an excess of electrons or negative charge at every point in the crystal where an As atom was introduced. This gives the crystal the n-type name. The opposite is true when a gallium atom with three valence electrons is used to dope a silicon crystal lattice. This creates a positive site or positive hole that is short one electron. The crystal is called a p-type semiconductor because of these sites with excess positive charge (shortage of negative charge).

7. The air in the frost-free refrigerator reaches an equilibrium between water molecules in the gas state and water molecules as frost. This equilibrium gas mixture is saturated with water molecules so it is "wet" or "moist". Dry air from the outside is blown into the refrigerator and the wet air is forced out. Then the gaseous water molecules and the frost (ice) reestablish equilibrium and the amount of frost decreases because some water molecules escape from the frost into the gas state. This process is repeated over and over again leading to the removal of frost in the refrigerator.

39. Superconductivity research is aimed at minimizing energy losses when transferring electricity at higher and higher temperatures. Present day superconducting transition temperatures are in the 70 to 100 kelvin range. This is expensive to reach and maintain on a large scale because of the cost of cryogenic liquids (liquid nitrogen) and handling equipment.

Solutions for Selected Problems

1. The first thing to do is to read the complete problem and decide what is being asked. This question asks for the new final volume. The second thing to do is to tabulate the information given.

   Balloon volume initial V_1 = 2205 mL   Balloon volume final V_2 = x
   Balloon pressure initial P_1 = 765 mmHg   Balloon pressure final P_2 = 725 mmHg

   Here all the volumes are in mL so there is no need for volume conversions. Similarly the pressures are both in mmHg so the units are consistent and no conversions are needed. Now decide what relationship to use. The calculations for all changes in state use some version of the combined gas law. In this problem the temperature is not mentioned so it is assumed to be constant. The combined gas law becomes Boyle's law.

   $$\text{Combined gas law} \quad \frac{P_1V_1}{T_1} = \frac{P_2V_2}{T_2} \qquad \text{Boyle's law} \quad P_1V_1 = P_2V_2$$

   Substituting for each quantity in Boyle's law (2090 mL)(765 mmHg) = (x)(725 mmHg)

   Solving for x; $x = \dfrac{(2090 \text{ mL})(765 \text{ mmHg})}{(725 \text{ mmHg})}$ = 2205 mL = 2210 mL rounded to 3 S.D.

   The answer is reasonable because experience shows that the volume increases when the pressure decreases. The final volume should be greater than the initial 2090 mL.

3. Reading the complete problem indicates that the question asks for the new or final pressure. Tabulating the information given shows that the volume units do not match. The volume units must either both be liters or else milliliters, mL. The conversion is shown for liters. P_1 = 6.25 mmHg P_2 = x

   $$V_1 = \text{45 Liter} \qquad V_2 = 745 \text{ mL} \times \frac{1 \text{ L}}{1000 \text{ mL}} = 0.745 \text{ L}$$

   Substituting into Boyle's law $P_1V_1 = P_2V_2$; (6.25 mmHg)(45 L) = (x)(0.745 L)

   Solving for $x = \dfrac{(6.25 \text{ mmHg})(45 \text{ L})}{(0.745 \text{ L})}$ = 378 mmHg = 380 mmHg rounded to 2 S. D.

   because the volume of 45 L has only 2 significant digits.

6. This question asks for the final volume after the pressure increases. Tabulate the information given in the question.

P_1 = 1.2 atm	P_2 = 2.4 atm
V_1 = 12 Liter	V_2 = x
T_1 = constant	T_2 = constant

    Boyle's law $P_1V_1 = P_2V_2$; Substituting into Boyle's law (1.2 atm)(12 L) = (2.4 atm)(x)

    Solving for x = $\dfrac{(12 \text{ liter})(1.2 \text{ atm})}{(2.4 \text{ atm})}$ = 6.0 L to 2 significant digits

    Doubling the pressure from 1.2 atm to 2.4 atm decreased the volume to one half what it was.

7. Conversion from one temperature scale to another may be made easier if you can use a scientific calculator. Some of them will do the conversion for you. There are reasons for not doing this problem that way. The logic of the conversion process is not visible when a push of a calculator button gives the answer. Another reason is that you probably don't have a calculator embedded in your arm so it won't always be available.

    a. The conversion of 22°C to kelvin uses the relation K = °C + 273.15;
        K = 22 + 273 = 295 K to the nearest whole number with three significant digits

    c. To convert 27.2°C to kelvins use K = 27.2 + 273.15 = 300.35 kelvin = 300.4 kelvins to one decimal place which would have four significant digits.

    e. To convert 29 kelvin to degrees Celsius use the relation K = °C + 273.15 solve for Celsius degrees. °C = K - 273.15;
        °C = 29 - 273.15 = -244 °C. to the nearest whole number

    f. To convert 631 kelvin to °C use the equation °C = K - 273.15;
        °C = 631 -273.15 = 358°C to the nearest whole number

8. The question asks for the new pressure and has constant volume. The problem uses the combined gas law simplified when volume stays constant.

    Combined gas law $\dfrac{P_1V_1}{T_1} = \dfrac{P_2V_2}{T_2}$ $\dfrac{P_1}{T_1} = \dfrac{P_2}{T_2}$

    Tabulate data. The temperatures must be converted to kelvins.

P_1 = 2.8 atm	P_2 = x
V_1 = constant	V_2 = constant
T_1 = -9 °C = 264 kelvin	T_2 = 28 °C = 301 kelvin

    Substituting in the pressure and temperature relation gives

    $\dfrac{2.8 \text{ atm}}{264 \text{ kelvin}} = \dfrac{x}{301 \text{ kelvin}}$ Solving for x = $\dfrac{(2.8 \text{ atm})(301 \text{ kelvin})}{264 \text{ kelvin}}$ = 3.19 atm = 3.2 atm

9. The question asks for the final temperature in kelvins and degrees Celsius.

Tabulating the data

P_1 = constant P_2 = constant

V_1 = 105 mL V_2 = 175 mL

$T_1 = 25^\circ C = 25 + 273 = 298$ kelvin T_2 = x

Combined gas law $\dfrac{P_1 V_1}{T_1} = \dfrac{P_2 V_2}{T_2}$ Charles law $\dfrac{V_1}{T_1} = \dfrac{V_2}{T_2}$

Substituting into Charles law $\dfrac{105 mL}{298 \text{ kelvin}} = \dfrac{175 mL}{x}$ Solving for x = $(298 \text{ kelvin})\dfrac{(175 mL)}{(105 \text{ mL})}$

$x = \dfrac{(298 \text{ kelvin})(175 \text{ mL})}{(105 \text{ mL})} = 497$ kelvin and 224 $^\circ$C

13. This question asks for the new volume. The tabulated data is given below. The two pressures are both in mmHg and need not be converted to other units. The Celsius temperatures must be converted to kelvins.

P_1 = 356 mmHg P_2 = 412 mmHg

V_1 = 350 mL V_2 = x

$T_1 = 23^\circ C + 273 = 296$ kelvin $T_2 = 21^\circ C + 273 = 294$ kelvin

The combined gas law is needed because no quantity stays constant and drops out.

Combined gas law $\dfrac{P_1 V_1}{T_1} = \dfrac{P_2 V_2}{T_2}$

Substituting into the combined gas law $\dfrac{(356 \text{ mmHg})(350 \text{ mL})}{(296 \text{ kelvin})} = \dfrac{(412 \text{ mmHg})(x)}{(294 \text{ kelvin})}$

Solve for the final volume. x = $\dfrac{(356 \text{ mmHg})(350 \text{ mL})(294 \text{ kelvin})}{(412 \text{ mmHg})(296 \text{ kelvin})} = 300$ mL

Speaking of Chemistry

Name _____

States of Matter

Across

2 Change from gas to liquid.
3 Solution that will not dissolve more solute.
5 Another name for the absolute temperature scale.
7 Abbreviation for kinetic molecular theory
9 Homogeneous mixtures of metals.
11 Weak electrostatic attraction between an electron rich atom and an hydrogen atom covalently bonded to a very electronegative atom like the H in one H_2O attracted to the O in another molecule.
18 Describes substances that are easily converted to gas from liquid.
19 The substance in a solution that is dissolved in a solvent.

Down

1 Solution that can dissolve more solute.

2 Gas law $\dfrac{V_1}{T_1} = \dfrac{V_2}{T_2}$
4 Physical change of a solid to liquid.
6 Semiconductor formed by doping a silicon with As atoms.
8 Device made from a layer of p-type silicon sandwiched between two n-type silicon layers.
10 Conversion of a liquid to a gas.
12 Gas law $P_1V_1 = P_2V_2$
13 Addition of a dopant like boron to silicon to make p-type silicon.
14 Another word for gas.
15 Elements that conduct electricity.
16 Substances like gases and liquids.
17 Semiconductors with positive "holes" for charge carriers, like B in silicon.

Bridging the Gap

Name _____

"Hot Molecules" Kinetic Molecular Theory

The kinetic molecular theory is sometimes difficult to visualize. This activity is aimed at showing how temperature influences the motion and therefore the kinetic energy of molecules. According to the kinetic theory of matter, the molecules of a substance are in constant motion. The molecules move faster at high temperatures and slower at low temperatures. This can be clearly seen by looking at the rates of diffusion of food coloring dye molecules in common tap water. The average kinetic energy of a molecule is given by the equation

$$\overline{KE} = \frac{1}{2}m\overline{v}^2 = \frac{3}{2}RT$$

Here the average kinetic energy is $\overline{KE}$, the molar mass is m, the average velocity for the molecules is $\overline{v}$, the universal gas constant is R, the absolute temperature is T. Remember the absolute temperature equals the Celsius temperature plus 273.15. This equation predicts that the average kinetic energy will increase with temperature. The physical result of higher velocities is a shorter time between molecular collisions. These collisions are important in diffusion and dispersion. Each collision will knock a molecule around a little bit. If a clump of dye molecules is buffeted by surrounding solvent molecules, each collision will act to break up the clump. Eventually the clump will be dispersed and the mixture will be uniform with a homogeneous color.

Equipment and materials

Microwave oven or source of hot water, ice cubes or source of ice cold water, three identical clear colorless 12 ounce glasses, plastic tumblers or jelly jars, three sheets of white paper, tap water and a package of Schilling® or other brand of food coloring.

1. Fill one of the glasses with approximately 8 ounces of room temperature water.

2. Fill the second glass with an equal amount of ice cold water.

3. Heat some water in either a microwave oven or on a stove top burner. Fill the third glass with an equal amount of hot water.

4. Place each of the glasses in front of a sheet of white paper so the paper acts as a white background.

5. Stir each glass of water and then allow the three glasses to stand for about 10 minutes so the water is as still as possible and there is no mechanical movement.

6. Select one of the food coloring dyes.

7. Carefully add one drop of dye to each of the glasses, starting with the coldest and ending with the hottest. Observe how the drop of dye spreads throughout the water.

8. Note the approximate time required for the food coloring to be uniformly dispersed in any one of the glasses.

Bridging the Gap
Concept Report Sheet
Name _____

"Hot Molecules" Kinetic Molecular Theory

Observations

What food coloring did you use?

Did the food coloring disperse at the same rate in the three different-temperature water samples? Which of the water samples dispersed the food coloring the fastest?

How did the drop of dye behave in the glass of room temperature water?

How did the drop of dye behave in the glass of hot water?

How did the drop of dye behave in the glass of ice cold water?

Analysis

What do you think causes the dye molecules in the food coloring to disperse the way they did in the three samples?

Do you believe you would observe the same results if you changed the food coloring? Justify your answer.

Do you believe you would see the same results if you used a different solvent like rubbing alcohol? Justify your answer.

Relation to Kinetic Molecular Theory

How do your observations relate to the Kinetic Molecular Theory?

Solution to Speaking of Chemistry crossword puzzle

States of Matter

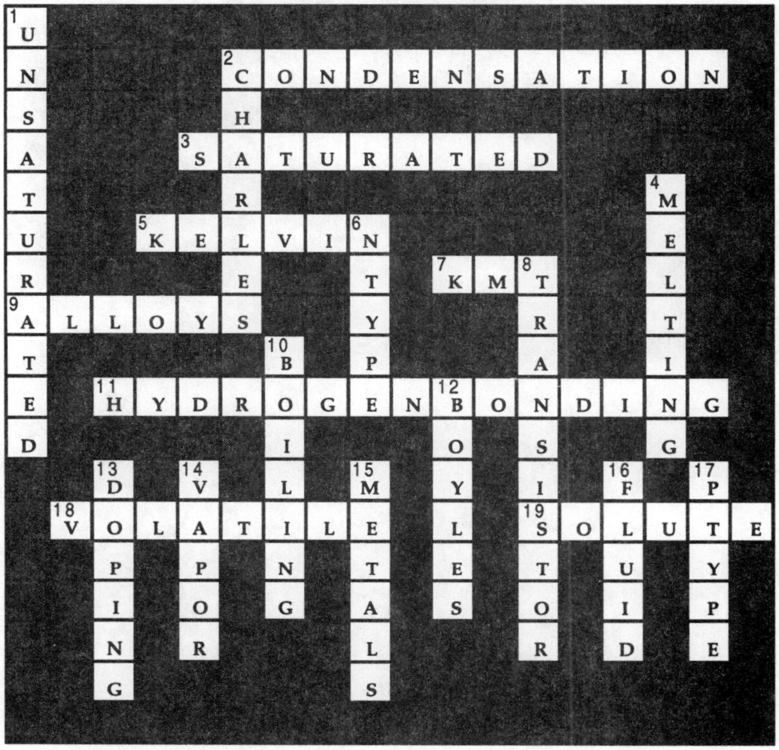
1 UNSATURATED
2 CONDENSATION
2 CHARLES
3 SATURATED
4 MELTING
5 KELVIN
6 NTYPE
7 KMT
8 TRANSISTOR
9 ALLOYS
10 BOILING
11 HYDROGEN BONDING
12 BOYLES
13 DOPING
14 VAPOR
15 METALS
16 FLUID
17 PTYPE
18 VOLATILE
19 SOLUTE

8 Chemical Reactivity: Chemicals in Action

8.1 Balanced Chemical Equations and What They Tell Us

This section describes how to read a chemical equation, and explains how the law of conservation of matter applies to chemical equations. The principles for balancing an equation are explained and illustrated with example reactions.

Objectives

After studying this section a student should be able to:

describe the characteristics of a balanced equation

describe the physical state of a reactant or product from subscript labels

identify reactants and products in an equation

distinguish a balanced equation from an unbalanced one

balance an equation by inspection given the formulas for reactants and products

explain why subscripts and formulas are not changed when balancing an equation

describe the role of coefficients in a balanced equation

identify the coefficients in a balanced equation

state the law of conservation of matter

Key terms

reactants	products	balanced equation	coefficients
physical state	(s)	(g)	(ℓ)
(aq)	subscripts	Conservation of Matter	

8.2 The Mighty Mole and the "How Much?" Question

This section explains and illustrates how to count objects by weighing. The mole is defined relative to Avogadro's number and related to the molar mass. Examples are given illustrating the computation of molar mass. The relationship between moles and chemical equations is explained and illustrated with examples.

Objectives

After studying this section a student should be able to:

describe the relationship between molar mass and Avogadro's number

give a definition for molar mass

show how to calculate the molar mass for a compound using the periodic table and formula for a compound

examine a chemical equation and identify the reactants and products

balance a chemical equation

tell what a diatomic molecule is and give some examples

Key terms

Avogadro's number	molar mass	mole
mass of reactant	mass of product	diatomic gas

8.3 Rates and Reaction Pathways: The "How Fast?" Question

Main topics in this section are the path a reaction takes, rates of reaction, factors influencing reaction rates, activation energy and its relationship to reaction rate. Reaction rates can be controlled by adjusting temperature, concentration of reactants and addition of a catalyst.

Objectives

After studying this section a student should be able to:

- state the relation between a reaction pathway and the rate of the process
- describe a reaction in terms of collisions between molecules
- state the role of a successful collision in the conversion of reactants to products
- give appropriate time units for slow and fast reaction rates
- state definitions for reaction rate, activation energy, reaction pathway
- describe how molecule kinetic energy can influence the success or failure of a collision
- tell what the effect of a catalyst is on the rate of a reaction and on the activation energy
- describe how the rate of a reaction is influenced by temperature and concentration
- explain what role enzymes play in reactions

Key terms

pathway	rate of a process	one-step
kinetic energy	kinetic molecular theory	successful collision
reaction rate	energy hill	fast reactions
slow reactions	collisions per second	activation energy
reaction direction	catalyst	effect of temperature
effect of concentration	alternate pathway	biological catalyst
enzymes	uncatalyzed reaction	

8.4   Chemical Equilibrium and the "How Far?" Question

This section explains what "dynamic equilibrium" means and describes chemical equilibrium for reversible reactions using the decomposition of $CaCO_3$. Changes in equilibrium conditions are explained using Le Chatelier's principle. Two different examples of equilibria are analyzed.

Objectives

After studying this section a student should be able to:

- give a definition for dynamic equilibrium
- describe an example of a dynamic equilibrium
- explain what a reversible reaction means
- state Le Chatelier's principle and give an example of how it works
- explain what happens when a reaction proceeds to completion

Key terms

dynamic equilibrium	chemical equilibrium	reversible
Le Chatelier's principle	forward reaction	reverse reaction
stress on a system	completion	

8.5   The Driving Forces and the "Why?" Question

This section describes potential energy and ties it to endothermic and exothermic reactions. The concept of favorable and unfavorable reactions is explained. The first law of thermodynamics is described using examples. The second law is explained in terms of entropy. Entropy is defined in terms of disorder. Entropy changes are linked to everyday events as well as chemical change.

Objectives

After studying this section a student should be able to:

- tell how a favorable reaction differs from an unfavorable reaction
- give the definition for endothermic reaction
- give the definition for exothermic reaction
- describe the relationship between entropy and disorder

104

tell how reactions relate to entropy increase and entropy decrease
explain how entropy content differs for solids, liquids, and gases
state the first law of thermodynamics
state the second law of thermodynamics

Key terms

favorable reaction	unfavorable reaction	potential energy
exothermic	endothermic	entropy
disorder	second law	first law of thermodynamics

8.6 More About the Mole and Chemical Reactions

This section shows how to use a balanced chemical equation to determine mole ratios for products and reactants and predict the number of moles needed or produced. The molar mass is combined with the mole ratio to determine grams of reactants and products.

Objectives

After studying this section a student should be able to:

identify a balanced equation
calculate the molar mass for a compound using the formula and the periodic table
examine a balanced equation and determine the mole ratio for any pair of substances
calculate the required number of moles of one reactant when given the moles of another and the balanced equation
calculate the number of moles product formed when given the number of moles of a reactant and the balanced equation
calculate the required grams of product formed when given the number of moles or grams of a reactant and the balanced equation
calculate the number of grams of one reactant required to combine with a given number of grams of another and the balanced equation

Key terms

mole	molar mass	mole ratio
completely react	moles of reactant	moles of product
mass of reactant	mass of product	

Additional Readings

Jones, Lynda. "Hot Gloves. (Spontaneous combustion by latex surgical gloves can cause fires)" Science World Nov. 3, 1995: 4.

Kingsbury, Donald. "The Janus-headed Arrow of Time: Entropy and Time Travel" Analog Science Fiction & Fact Feb. 1995: 58.

Zibel, Alan. "Recombinant Detergent" Popular Science Jan. 1996: 35.

Answers to Odd Numbered Questions for Review and Thought

1. a. $CH_3CH_2OH_{(g)} + 3 O_{2(g)} ---> 2 CO_{2(g)} + 3 H_2O_{(g)}$
    $1 \times (3 + 2 + 1) = 6$ H atoms $3 \times (2) = 6$ H atoms
    In one molecule of CH_3CH_2OH or $CH_3CH_3O_1$ on the reactant side of the equation, there are 6 H atoms; this number is found by adding together the subscripts for H in the formula and multiplying by the coefficient, 1 (understood), in front of the ethanol formula in the balanced equation. The product side of the balanced equation also shows 6 H atoms; there are three water molecules on this side and each molecule

contains 2 H atoms so the number of H's is found by multiplying the coefficient (3) by the subscript (2) for H in the water formula.

b. $CH_3CH_2OH_{(g)}$ + 3 $O_{2(g)}$ ---> 2 $CO_{2(g)}$ + 3 $H_2O_{(g)}$
Each side of this balanced equation shows 7 oxygen atoms, O. On the left there are 3 oxygen molecules, O_2, and each oxygen molecule has 2 oxygen atoms in it; there is one oxygen atom in the CH_3CH_2OH molecule; $3 \times 2 + 1 = 7$. On the right or product side there are 2 molecules of CO_2, and each contains 2 oxygen atoms; this accounts for 4 oxygen atoms. There are also 3 water molecules, each containing 1 oxygen atom; this accounts for 3 more O atoms. Therefore, on the right side of the equation there are $4 + 3 = 7$ oxygen atoms shown.

c. $CH_3CH_2OH_{(g)}$ + 3 $O_{2(g)}$ ---> 2 $CO_{2(g)}$ + 3 $H_2O_{(g)}$
The balancing coefficients are 1, 3, 2, and 3, respectively. The "1" in front of CH_3CH_2OH is understood to be there even though it usually is not written.

d. The count for the number of atoms of an element is the same for both reactant and product sides of the equation. For example, in the equation above, there are 6 H atoms on the left of the arrow and 6 H atoms on the right side; there are 7 O atoms on each side and also 2 C atoms on each side.

3. a. To balance the equation Al + Cl_2 ---> $AlCl_3$ notice that the subscripts on the Cl are 2 and 3 on the left and right sides, respectively, of the arrow. Neither is a factor of the other , but this pair has a least common multiple of 6. The coefficient in front of Cl_2 must be "3" to give 6 Cl atoms on the left. The coefficient for $AlCl_3$ must be "2" to give 6 Cl atoms on the right. Al + 3 Cl_2 ---> 2 $AlCl_3$ Now turn to the aluminum, Al. On the right the count for Al is 2 , because of the coefficient "2" in front of the $AlCl_3$. This means that 2 Al's must appear on the left side as well, so we need to put a "2" in front of the Al. The complete balanced equation is:
$$2\ Al\ +\ 3\ Cl_2\ --->\ 2\ AlCl_3.$$

b. Mg + N_2 ---> Mg_3N_2 Notice that the subscript for N is a 2 in both reactant and product. This means that the coefficient for the N_2 will be the same as the coefficient for Mg_3N_2; the smallest possible coefficient is "1". $Mg + 1\ N_2$ ---> $1\ Mg_3N_2$ Now count the magnesium atoms. The product has $1 \times 3 = 3$ Mg atoms. To equal this we put a "3" in front of the Mg on the left of the equation.
$$3\ Mg\ +\ N_2\ --->\ Mg_3N_2$$

c. NO + O_2 ---> NO_2 Examine the equation and count the number of O atoms on each side and the number of N atoms on each. There are 2 oxygen atoms in NO_2 so whatever coefficient we place in front of this in the equation will give us an even number of oxygen atoms on the right; the number of oxygen atoms on the right will be 2 or 4 or 6, etc. The left side already shows $1 + 2 = 3$ oxygen atoms so we know we can't get a balanced equation with only 2 oxygens on the right. This leads us to try a "2" in front of the NO_2 to give 4 oxygen atoms and 2 nitrogen atoms on the right. To get 2 nitrogen atoms on the left we need a "2" in front of the NO; this will also account for 2 oxygen atoms in the two molecules of NO. Adding the 2 oxygen atoms shown in the 2 NO's to the 2 oxygen atoms shown in one O_2 molecule gives 4 oxygen atoms on the left.
$$2\ NO\ +\ 1\ O_2\ --->\ 2\ NO_2$$

106

d. $SO_2 + O_2 \longrightarrow SO_3$ The SO_2 and SO_3 molecules each contain one S atom so we know the coefficients for SO_2 and SO_3 must be the same. We also know that a "1" will not work for this coefficient because that would give 3 oxygen atoms on the right and we can already see 4 oxygens on the left. So, if the coefficient "1" will not work, we next try a "2". Trying a "2" in front of SO_3 and a "2" in front of SO_2 balances the S atoms. Next, looking at the oxygen atoms we see 2 x 3 or 6 oxygen atoms on the right; we see 2 x 2 oxygens in the 2 SO_2's on the left and another 2 oxygens in the single O_2 molecule, so we can account for 6 oxygen atoms on the left.

$$2\ SO_2 + O_2 \longrightarrow 2\ SO_3$$

e. $H_2 + N_2 \longrightarrow NH_3$ This equation is balanced by seeing that the subscripts for hydrogen are a 2 in the H_2 and a 3 in the ammonia, NH_3. The lowest common multiple of 2 and 3 is 6 so the coefficient for the H_2 must be a "3" and the coefficient for the NH_3 must be a "2". Next, checking the nitrogen atoms we see 2 N's in the 2 NH_3 molecules on the right. For 2 N's on the left we need one N_2 molecule

$$3\ H_2 + N_2 \longrightarrow 2\ NH_3$$

5. a. $Ba_{(s)} + H_2O_{(l)} \longrightarrow Ba(OH)_{2(aq)} + H_{2(g)}$ This reaction is balanced more easily when the break up of the H_2O molecule into OH and H is recognized. This often happens when water is a reactant. This means that each OH among the products came from one water molecule or that every OH in products must be matched by an H_2O molecule in the reactants. To get 2 OH units in the $Ba(OH)_2$ product requires that 2 H_2O molecules react. If 2 water molecules break apart, there will also be 2 H's produced that can join to make one H_2 molecule. The coefficients for Ba and $Ba(OH)_2$ must be the same because the subscripts for Ba are the same in both.

$$Ba_{(s)} + 2\ H_2O_{(l)} \longrightarrow Ba(OH)_{2(aq)} + H_{2(g)}$$

b. $Fe_{(s)} + H_2O_{(l)} \longrightarrow Fe_3O_{4(s)} + H_{2(g)}$ Note that the oxygen atoms in the Fe_3O_4 product must come from the H_2O molecules on the left. To get 4 oxygens on the right we pick a coefficient of "4" to place in front of the H_2O.

$Fe_{(s)} + 4\ H_2O_{(l)} \longrightarrow Fe_3O_{4(s)} + H_{2(g)}$ The subscript for the H is the same in both H_2O and H_2 and hydrogen appears only in these two molecules, so the coefficients for H_2O and H_2 must be the same (4). Alternatively, we could look at the 4 H_2O on the left and recognize that there are 8 H's there. To get 8 H's on the right requires a coefficient of "4" in front of the H_2. Lastly, there must be 3 Fe's on each side so we put a "3" in front of the Fe on the left.

$$3\ Fe_{(s)} + 4\ H_2O_{(l)} \longrightarrow Fe_3O_{4(s)} + 4\ H_{2(g)}$$

c. $Na_{(s)} + H_2O_{(l)} \longrightarrow NaOH_{(aq)} + H_{2(g)}$ This reaction is similar to 5a. The H_2O molecules break apart to yield OH and H. It is helpful to notice that the coefficients for Na and NaOH must be the same because one Na is shown is each; the coefficients for H_2O and NaOH must be the same because the subscripts on the oxygen are the same; the coefficients for H_2O and H_2 are the same because the subscripts on H are the same. If we recognize that one H comes from the break up of one water molecule, then we can conclude that 2 water molecules are needed to give up the 2 H's shown in the H_2 molecule. We put a "2" in front of the H_2O. This will also produce 2 OH's so we need a "2" in front of the NaOH to account for 2 OH's. This then shows 2 Na's on the right and means we must put a "2" in front of the Na on the left.

$$2\ Na_{(s)} + 2\ H_2O_{(l)} \longrightarrow 2\ NaOH_{(aq)} + H_{2(g)}$$

d. This equation is balanced like 5 c. except that Li is reacting in place of Na. This shows how all the elements of Group IA react with water.

$$2 \; Li_{(s)} \; + \; 2 \; H_2O_{(l)} \; \text{---> } \; 2 \; LiOH_{(aq)} \; + \; H_{2(g)}$$

7. a. $Sn_{(s)} \; + \; HBr_{(aq)} \; \text{---> } SnBr_{2(aq)} \; + \; H_{2(g)}$ Note that the subscript for Br in HBr is a 1 and the subscript for Br in $SnBr_2$ is a 2. The lowest common multiple is $2 \times 1 = 2$; the coefficient for the HBr must be a "2" if the coefficient for $SnBr_2$ is a "1".

$$Sn_{(s)} \; + \; 2 \; HBr_{(aq)} \; \text{---> } SnBr_{2(aq)} \; + \; H_{2(g)}$$

This also shows one Sn and 2 H's on each side of the arrow so the equation is balanced.

b. $Mg_{(s)} \; + \; HCl_{(aq)} \; \text{---> } \; MgCl_{2(aq)} \; + \; H_{2(g)}$ This equation is balanced the same way as 7 a. except that Mg is reacting instead of Sn and HCl is used in place of HBr. Note that all Group IIA metals will react in a similar way with HCl.

$$Mg_{(s)} \; + \; 2 \; HCl_{(aq)} \; \text{---> } \; MgCl_{2(aq)} \; + \; H_{2(g)}$$

c. This equation is balanced in the same way as 5 a. Ba and Ca are both from Group IIA and react similarly. $Ca_{(s)} \; + \; 2 \; H_2O_{(l)} \; \text{---> } \; 2 \; Ca(OH)_{2(aq)} \; + \; H_{2(g)}$

d. $Zn_{(s)} \; + \; HNO_{3(aq)} \; \text{---> } \; Zn(NO_3)_{2(aq)} \; + \; H_{2(g)}$ This equation is balanced more easily if we recognize that the HNO_3 molecule comes apart into NO_3 and H, much as the Cl and H separate in the HCl in 7 a. To find 2 NO_3 units in the product $Zn(NO_3)_2$ requires the break up of 2 HNO_3's, so we put a "2" in front of the HNO_3 on the left. Break up of 2 HNO_3's also frees 2 H's which can go together to make one H_2 molecule. Since there is one Zn on the right, we only need one Zn on the left.

$$Zn_{(s)} \; + \; 2 \; HNO_{3(aq)} \; \text{---> } \; Zn(NO_3)_{2(aq)} \; + \; H_{2(g)}$$

e. This equation is balanced just like 5 c. and 5 d. except that cesium, Cs, is the reactant metal instead of Na or Li. Note that Li, Na, and Cs are all members of Group IA in the periodic table. $2 \; Cs_{(s)} \; + \; 2 \; H_2O_{(l)} \; \text{---> } \; 2 \; CsOH_{(aq)} \; + \; H_{2(g)}$

9. The carbon-12 isotope with a _defined_ mass of <u>exactly</u> 12 amu is the basis. The element carbon has an atomic weight greater than 12.0000. because some naturally-occurring isotopes of carbon are heavier than the carbon-12 atoms; this makes the atomic weight larger than 12 amu since atomic weight shows the weight of an *average* carbon atom.

11. Each is 0.500 mol of that element. 1 mol Pb = 207.0 g Pb and 1 mol C = 12.012 g C.

$$\frac{103.5 \; g \; Pb}{1} \; \times \; \frac{1 \; mol \; Pb}{207.0 \; g \; Pb} = 0.5000 \; mol \; Pb \qquad \frac{6.006 \; g \; C}{1} \; \times \; \frac{1 \; mol \; C}{12.012 \; g \; C} = 0.5000 \; mol \; C$$

13. Avogadro's number is 6.022×10^{23} to 4 significant digits or 6.02×10^{23} to 3 s.d.

15. The mol is defined as the number of atoms in exactly 12 g of carbon-12, about 6.02×10^{23} atoms.

108

17. Remember that 1 mol (or mole) = 6.02×10^{23} formula units. These definitions can be used as conversion factors to calculate number of formula units when we are given the number of mols of element or compound.

   a. 1 mol CO_2 = 6.02×10^{23} CO_2 molecules

   b. $\dfrac{0.001 \text{ mol He}}{1} \times \dfrac{6.02 \times 10^{23} \text{ He atoms}}{1 \text{ mol He}} = .00602 \times 10^{23} = 6.02 \times 10^{20}$ He atoms

   c. $\dfrac{10 \text{ mol butane}}{1} \times \dfrac{6.02 \times 10^{23} \text{ butane molecules}}{1 \text{ mol butane}} = 60.2 \times 10^{23} = 6.02 \times 10^{24}$ butane molecules

   d. $\dfrac{32 \text{ mol } O_2}{1} \times \dfrac{6.02 \times 10^{23} \text{ } O_2 \text{ molecules}}{1 \text{ mol } O_2} = 193 \times 10^{23} = 1.9 \times 10^{25}$ O_2 molecules

   e. $\dfrac{0.001 \text{ mol He}}{1} \times \dfrac{6.02 \times 10^{23} \text{ He atoms}}{1 \text{ mol He}} = .00602 \times 10^{23} = 6.02 \times 10^{20}$ He atoms

19. Some reactions are fast because reactants have weak bonds. These reactions have low activation energies. (These reactions have a low energy barrier or a "low hill to climb" for reactants to form products.) Reactions between gas phase molecules occur rapidly because collisions between reactant molecules are more frequent for gases. Some reactions are slow because reactants have strong bonds and a high activation energy. These reactants have a high "energy hill to climb" to form products.

21. Freezing slows reaction rates; this slows spoilage and slows the growth of mold or bacteria. The lower temperature decreases the kinetic energy of reactant molecules, causing them to move more slowly. This decreases the chance of a collision between reactants. Also the collisions will have low energies when they do occur, so the chance of having the necessary activation energy is less. The freezing process tends to immobilize molecules in a solid where the freedom of movement is very low, diminishing chances for collisions even more. This means that molds and bacteria grow much more slowly.

23. Reactions with high or large activation energies (examples I and II) are slower than reactions with low or small activation energies (examples III and IV).

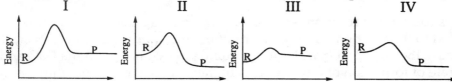

25. The activation energy is high (large) so there is no reaction at room temperature. The spark provides the needed activation energy to start the reaction. The reaction is exothermic and the energy released provides the energy needed to keep the reaction going; the reaction occurs quickly since it has a "built-in" source of energy. Any "extra" energy is released to the surroundings. A reaction produces energy because the products have lower potential energy than the reactants.

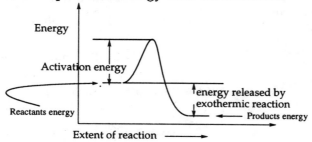

27. Burning paper requires that the molecules in the paper react with O_2 molecules. Paper (and other substances) burn more rapidly in pure oxygen because collisions between oxygen molecules and the paper are more numerous or more frequent if oxygen is the only gas present. Air is about 80% nitrogen gas and only about 20% oxygen gas, so in air only about 20% of the collisions between the paper and the gas molecules will involve an O_2 molecule and contribute to the reaction. When N_2 molecules collide with the paper, there will be no reaction.

29. Reversible reactions occur in both forward and reverse directions. Consider the reaction $N_2 + 3H_2 \rightleftharpoons 2NH_3$ The double arrow indicates reversibility. This means that both the reaction $N_2 + 3H_2 \longrightarrow 2NH_3$ and the reaction $2NH_3 \longrightarrow N_2 + 3H_2$ can and do occur at the same time. When the reaction rates for both of these reactions are exactly the same, equilibrium exists between the forward and reverse processes. As fast as 2 ammonia molecules form, somewhere else in the reaction mixture 2 ammonia molecules are being converted to 1 nitrogen and 3 hydrogen molecules. Once equilibrium is established between forward and reverse processes, the net amount of each substance in the reaction mixture stays the same.

31. When limestone is roasted the reaction is $CaCO_{3(s)} \rightleftharpoons CaO_{(s)} + CO_{2(g)}$
    a. If CO_2 is added to the reaction mixture, the reaction will shift toward $CaCO_{3(s)}$. This happens because the added CO_2 will initially collide more often with the CaO and increase the rate of formation of $CaCO_3$. This higher rate will continue until a new balance for equilibrium comes into being. The new equilibrium will involve slightly more CO_2, less CaO, and more $CaCO_3$ than the original equilibrium balance.
    b. If carbon dioxide is allowed to escape, the reaction will shift to replace the missing CO_2. This means that the rate of disappearance of $CaCO_{3(s)}$ will initially increase as CO_2 and CaO form until a new equilibrium is established. The new equilibrium will contain less $CaCO_3$, more CaO and slightly less CO_2 that the original.

33. When a reversible reaction shifts to favor products, additional reactants are consumed to make more products until a new balance is reached.

110

35. After the reaction of HCl and NaOH has gone to completion, the reactant concentrations are practically zero; essentially, all reactants have been converted to products.

37. a. Potential energy is stored energy or energy due to the positions of particles. A book on a shelf has potential energy relative to the floor below. Two charged particles have potential energy due to the attraction (opposite charges) or repulsion (like charges) between them.

   b An exothermic process gives off energy.
   c. An endothermic process consumes or takes in energy.
   d. Entropy is a measure of chaos or disorder. A well-ordered deck of cards has low entropy and a shuffled deck has higher disorder and entropy. Natural processes tend to go to higher entropy. Therefore, energy and work must be used to reduce the entropy of these systems.
   e. A favorable reaction is a process that favors products at the expense of reactants. The reaction is described as going to completion.

39. You can write a book answering this question! The symbol for entropy is "S". A change is entropy is represented by ΔS. Every spontaneous process occurs with an increase in entropy for the universe. The total entropy change for the universe is the sum of the entropy change for the system being examined plus the entropy change for the surroundings: $\Delta S_{universe} = \Delta S_{system} + \Delta S_{surroundings}$. This entropy change for the universe consumes energy that cannot be used for other purposes. Every spontaneous process has a built-in amount of so called "wasted" energy.

41. Burning methane is an exothermic reaction. We observe this when we notice that burning methane or natural gas produces heat: $CH_4 + 2 O_2 ---> CO_2 + 2 H_2O + heat$.

43. No, it is not favorable because it requires a continuous input of energy to keep the process going; without this energy the process will stop. The entropy decrease for building up a large molecule from several small ones is accompanied by a larger entropy increase in the surroundings. The total entropy of the universe increases:
$\Delta S_{universe} = \Delta S_{photosynthesis} + \Delta S_{surroundings} = +$ value. The chaos created in the surroundings exceeds the organization created in making the large glucose molecule.
$6 CO_2 + 6 H_2O ---> C_6H_{12}O_6 + O_2$

45. No, recycling of waste does not violate the 2nd law by creating order in the system of waste material. The energy used to organize the recycled material creates disorder in the surroundings. The resulting total entropy change is an increase in disorder. This is illustrated by the production of entropy when people use energy to sort and collect recycleables. $\Delta S_{universe} = \Delta S_{recycled\ waste} + \Delta S_{sorting,\ collecting,\ processing}$

Solutions for Selected Problems

1. The molar mass is calculated using the formulas and the Atomic Weights from the periodic table. The first step is to count the number of atoms of each element in the formula; this tells the number of mols of each element needed for one mol of the compound. Masses are rounded to give whole numbers for ease of calculation; in other problems masses will be rounded to the number of significant digits appropriate to match the factors or terms in the rest of the problem.

a. Atomic Weight of H = 1.01. Atomic Weight of O = 16.00
One molecule of water contains 2 atoms of hydrogen and 1 atom of oxygen.
One *mol* of water contains 2 *mols* of hydrogen and 1 *mol* of oxygen.

$$2 \text{ mol H} = 2 \text{ g H} \qquad\qquad 2 \text{ mol H} \times \frac{1.01 \text{ g H}}{1 \text{ mol H}} = 2.02 \text{ g H} = 2 \text{ g H}$$

$$1 \text{ mol O} = 16 \text{ g O} \qquad\qquad 1 \text{ mol O} \times \frac{16.00 \text{ g O}}{1 \text{ mol O}} = 16.00 \text{ g O} = 16 \text{ g O}$$

$$1 \text{ mol H}_2\text{O} = 2 + 16 = 18 \text{ g H}_2\text{O}$$

b. 1 molecule I_2 contains 2 atoms I; 1 mol I_2 contains 2 mols I. Atomic Weight I = 127

$$\text{molar mass } I_2 = \text{ mass of 1 mol of } I_2 = 2 \text{ mol I} \times \frac{127 \text{ g I}}{1 \text{ mol I}} = 254 \text{ g } I_2$$

c. 1 mol KOH contains 1 mol K, 1 mol O and 1 mol H.
molar mass KOH = 39 g K + 16 g O + 1 g H = 56 g KOH

d. 1 mol NH_3 contains 1 mol N and 3 mols H
molar mass NH_3 = 14 g N + (3 mols H)(1 g H/ 1 mol H) = 17 g NH_3

e. 1 mol CO_2 contains 1 mol C and 2 mols O
molar mass CO_2 = 12 g C + (2 mol O)(16 g O/ 1 mol O) = 12 + 32 = 44 g CO_2

f. 1 mol CO contains 1 mol C and 1 mol O
molar mass CO = 12 g C + 16 gO = 28 g CO

3. Look up the Atomic Weight of each element in the periodic table. This also tells the mass in grams of 1 mol of the element. "Molar mass" means the mass of 1 mol of substance.
a. 1 mol Cu = 63.55 g Cu;   b. 1 mol Pb = 207.2 g Pb;   c. 1 mol Na = 22.99 g Na.
The molar mass of lead, Pb, is the largest .

5. We must calculate the molar mass and then use it as a conversion factor.
a. 1 mol C_4H_{10} contains 4 mols C and 10 mols H

$$4 \text{ mol C} \times \frac{12 \text{ g C}}{1 \text{ mol C}} = 48 \text{ g C} \qquad 10 \text{ mol H} \times \frac{1 \text{ g H}}{1 \text{ mol H}} = 10 \text{ g H}$$

molar mass C_4H_{10} = 48 g C + 10 g H = 58 g C_4H_{10} = mass of 1 mol C_4H_{10}

$$58 \text{ g } C_4H_{10} \times \frac{1 \text{ mol } C_4H_{10}}{58 \text{ g } C_4H_{10}} = 1.0 \text{ mol } C_4H_{10}$$

b. 1 mol Mg(OH)$_2$ contains 1 mol Mg, 2 mols O and 2 mols H
molar mass Mg(OH)$_2$ = 24.31 g Mg + (2)(16.00) g O + (2)(1.01) g H = 58.33 g Mg(OH)$_2$

$$.63 \text{ g Mg(OH)}_2 \times \frac{1 \text{ mol Mg(OH)}_2}{58.33 \text{ g Mg(OH)}_2} = 0.11 \text{ mol Mg(OH)}_2$$

c. molar mass Na = 22.99 g

$$230 \text{ g Na} \times \frac{1 \text{ mol Na}}{22{:}99 \text{ g Na}} = 10. \text{ mol Na to 2 significant digits}$$

6. The conversion of mols of substance to grams of substance also requires the molar mass of the substance as a conversion factor. Calculate the molar mass of the substance then convert to grams of substance.

a. molar mass NH$_3$ = 14.01g N + 3.03 g H = 17.04 g NH$_3$; 2 significant digits are enough to go with the 2 significant digits in "4.5 mols ammonia"

$$4.5 \text{ mols NH}_3 \times \frac{17 \text{ g NH}_3}{1 \text{ mol NH}_3} = 76.5 \text{ g NH}_3 = 77 \text{ g NH}_3 \text{ to 2 significant digits}$$

b. 1 mol H$_2$O$_2$ = 2 mol H + 2 mol O = (2)(1.0) g H + (2)(16.0) g O = 34.0 g H$_2$O$_2$

$$0.0023 \text{ mols H}_2\text{O}_2 \times \frac{34.0 \text{ g H}_2\text{O}_2}{1 \text{ mol H}_2\text{O}_2} = 0.0782 \text{ g H}_2\text{O}_2 = 0.078 \text{ g H}_2\text{O}_2 \text{ to 2 significant digits}$$

7. Write the balanced equation and include the molar interpretation, i.e. read the balanced equation in terms of mols of reactants and products.

$$\begin{array}{ccc}
\text{H}_{2(g)} & + \quad \text{Br}_{2(\ell)} ---> & 2\text{ HBr}_{(g)} \\
1 \text{ mol H}_2 & 1 \text{ mol Br}_2 & 2 \text{ mols HBr}
\end{array}$$

Calculate the molar mass for each substance in the equation.
1 mol H$_2$ = 2.02 g H$_2$; 1 mol Br$_2$ = 159.8 g Br$_2$; 1 mol HBr = 80.9 g HBr
Replace molar interpretation with molar masses for each entry in the equation and compare these with the masses given in the problem. Predict the mass of product using the law of conservation of mass.

$$\begin{array}{ccc}
\text{H}_{2(g)} \ + & \text{Br}_{2(\ell)} ---> & 2\text{ HBr}_{(g)} \\
1 \text{ mol H}_2 & 1 \text{ mol Br}_2 & 2 \text{ mols HBr} \\
2.02 \text{ gH}_2 & 159.8 \text{ g Br}_2 & (2)(80.9) \text{ g HBr} = 161.8 \text{ g HBr}
\end{array}$$

The masses given in the problem are:
2.0 g H$_2$ 159.8 g Br$_2$ 161.8 g HBr is predicted as the product mass

9. First step is to balance the equation:

$$C_6H_{12}O_{6(aq)} \longrightarrow 2\ CH_3CH_2OH_{(aq)} + 2\ CO_{2(g)}$$

The molar interpretation can be made from the coefficients in the balanced equation.

$C_6H_{12}O_{6(aq)}$	$\longrightarrow$	$2\ CH_3CH_2OH_{(aq)}$	$+$	$2\ CO_{2(g)}$
1 mol $C_6H_{12}O_6$		2 mols CH_3CH_2OH		2 mols CO_2

a.   There are two ways to approach the solution to this problem. We can consider what happens to the amounts of products if the amount of glacés is 6 times as much as the amount of glucose shown in the balanced equation; if 6 mols of glucose react, the amounts of ethanol and carbon dioxide will also be multiplied by 6.

glucose		ethanol		carbon dioxide
$C_6H_{12}O_{6(aq)}$	$\longrightarrow$	$2\ CH_3CH_2OH_{(aq)}$	$+$	$2\ CO_{2(g)}$
1 mol $C_6H_{12}O_6$		2 mols CH_3CH_2OH		2 mols CO_2
6 x 1 mol $C_6H_{12}O_6$		6 x 2 mols CH_3CH_2OH		6 x 2 mols CO_2
6 mol $C_6H_{12}O_6$		12 mols CH_3CH_2OH		12 mols CO_2

The short method to solve this problem is to determine the mol-ratio between glucose and ethanol from the balanced equation and use this mol-ratio as a conversion factor.

$$\frac{1\ mol\ C_6H_{12}O_6}{2\ mol\ CH_3CH_2OH} \quad or \quad \frac{2\ mol\ CH_3CH_2OH}{1\ mol\ C_6H_{12}O_6}$$

$$6\ mols\ C_6H_{12}O_6 \times \frac{2\ mol\ CH_3CH_2OH}{1\ mol\ C_6H_{12}O_6} = 12\ mols\ CH_3CH_2OH$$

b.   The mols of ethanol can be determined from mols of carbon dioxide using the mol-ratio that comes from the balanced equation: 2 mols ethanol/ 2 mols carbon dioxide.

$$10.5\ mols\ CO_2 \times \frac{2\ mol\ CH_3CH_2OH}{2\ mol\ CO_2} = 10.5\ mols\ CH_3CH_2OH$$

11. a. This problem is similar to example 8.7. Steps to follow to solve the problem are:

1. Balance the equation
2. Determine mol-ratio needed by giving the molar interpretation of the equation.
3. Calculate molar masses for substances involved in the problem
4. Convert "grams Sn" to "mols Sn" using its molar mass.
5. Convert "mols Sn" to "mols SnO" using the mol-ratio from the balanced equation.
6. Convert "mols SnO" to "grams SnO" or "g SnO" using the molar mass for SnO.

1. $2\ Sn \quad + \quad O_2 \quad ---> \quad 2\ SnO$

2. 2 mols Sn   2 mols SnO

3. molar mass Sn = 119 g Sn and molar mass SnO = 119 g Sn + 16 g O = 135 g SnO

4. $11.9\ g\ Sn \times \dfrac{1\ mol\ Sn}{119\ g\ Sn} = 0.100\ mols\ Sn$

5. $11.9\ g\ Sn \times \dfrac{1\ mol\ Sn}{119\ g\ Sn} \times \dfrac{2\ mol\ SnO}{2\ mol\ Sn} = 0.100\ mol\ SnO$

6. $11.9\ g\ Sn \times \dfrac{1\ mol\ Sn}{119\ g\ Sn} \times \dfrac{2\ mol\ SnO}{2\ mol\ Sn} \times \dfrac{135\ g\ SnO}{1\ mol\ SnO} = 13.5\ g\ SnO$

Cancel the units in the set up to check the problem and make sure the answer is "g SnO".

b. Follow a series of steps similar to the 6 steps above to work this problem.

1. $2\ Sn \quad + \quad O_2 \quad ---> \quad 2\ SnO$
2. 2 mols Sn   1 mol O_2
3. molar mass Sn = 119 g Sn and molar mass O_2 = 32.00 g O_2
Steps 4, 5, and 6 can be combined into one set up.

4.-6. $5.4\ g\ Sn \times \underset{\text{step 4.}}{\dfrac{1\ mol\ Sn}{119\ g\ Sn}} \times \underset{\text{step 5.}}{\dfrac{1\ mol\ O_2}{2\ mol\ Sn}} \times \underset{\text{step 6.}}{\dfrac{32.00\ g\ O_2}{1\ mol\ O_2}} = 0.72\ g\ O_2$

c. Follow a series of steps similar to the 6 steps in part "a." to work this problem.

1. $2\ Sn \quad + \quad O_2 \quad ---> \quad 2\ SnO$
2. 2 mols Sn   2 mols SnO
3. molar mass Sn = 119 g Sn and molar mass SnO = 119 g Sn + 16 g O = 135 g SnO
Steps 4, 5, and 6 can be combined into one set up.

4.-6. $5.4\ g\ Sn \times \underset{\text{step 4.}}{\dfrac{1\ mol\ Sn}{119\ g\ Sn}} \times \underset{\text{step 5.}}{\dfrac{2\ mol\ SnO}{2\ mol\ Sn}} \times \underset{\text{step 6.}}{\dfrac{135\ g\ SnO}{1\ mol\ SnO}} = 6.13\ g\ SnO$

Speaking of Chemistry

Name _____

Chemical Reactivity

Across

2 Process that releases heat.
4 Change applied to an equilibrium system.
6 The _____ law of thermodynamics, predicts conservation of energy.
8 A biological catalyst.
11 The strength of solutions and mixtures are measured in _____ units.
12 The speed of reaction and change of concentration with time indicate the _____ .
13 Reaction rate is_____when reactants are consumed quickly.
14 The number 2 and 3 in the Fe_2O_3 formula are called_____.
15 Process that proceeds in both the forward and reverse directions.
16 An Avogadro's number of formulas.

Down

1 State in which particles are in contact and in fixed positions.
3 Materials on the left hand or beginning side of an equation
5 The _____ law of thermodynamics says entropy is always increasing.
7 A _____ increase will raise the kinetic energy of particles.
9 Stored energy.
10 Physical state where particles are far apart.

Bridging the Gap I Name _____

Sodium Bicarbonate, NaHCO3, and Vinegar
Heat Effects and Production of CO2

This exercise will deal with the reaction between vinegar and baking soda. Vinegar is a mixture that is roughly 95% water and 5 % acetic acid, CH_3COOH. Baking soda is pure sodium bicarbonate, $NaHCO_3$. The reaction is shown below.

Acetic acid   Baking soda   Water   Carbon dioxide Sodium acetate

$CH_3COOH(aq)$ + $NaHCO_3(s)$ $\longrightarrow$ $H_2O(l)$ + $CO_2(g)$ + $Na^+CH_3COO^-(aq)$

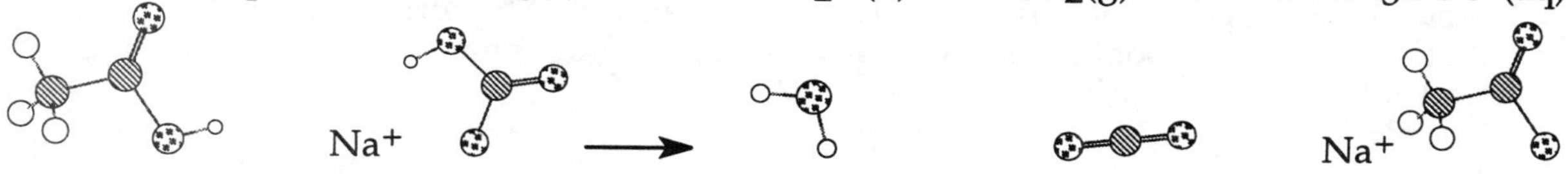

Na^+

Na^+

Heat Effects

One part of this activity is to observe the heat effect that accompanies the reaction. This will allow you to decide if the reaction is exothermic or endothermic. When a reaction is exothermic the reaction releases energy in the form of heat or light to the surroundings. A fire is a reaction that produces both heat and light. Sometimes this energy release is only in the form of heat. The surroundings gain energy (from the reaction) and become warmer. You can observe this by touching the container. When a reaction is endothermic the reaction draws energy from the surroundings. In this case the surroundings lose energy and create a "cooling" sensation if you touch the reaction container.

Production of Carbon Dioxide, CO2(g)

This part of the exercise is to observe the reaction between vinegar and baking soda and note the amount of gas, CO_2, produced.

Equipment and materials

A box of Ziploc® sandwich bags (or equivalent), a roll of Saran® or other plastic wrap, a bottle of plain white vinegar, a box of baking soda (typically Arm and Hammer®), a 1/4 cup measuring cup, 1/8 and 1/4 teaspoon measuring spoons, scissors, marking pen, and transparent tape or stick on labels.

Procedure

1. Label three Ziploc® bags as 1 , 2, and 3. Fold back the top of the open Ziploc® bag back so the bag stay open. Do this with all three bags. Set the bags in a bowl or cup so they stand up and don't spill.

2. Add 1/4 cup of vinegar to each one of the three Ziploc® bags.

3. Open the box of baking soda and stir the contents so that samples are not taken only from the surface which may have reacted or decomposed.

4. Cut three pieces of plastic wrap about 3x3 inches each. These will be used to wrap the sodium bicarbonate (baking soda).

5 Layout the three pieces of plastic wrap on a clean surface. Measure out the baking soda with the measuring spoon. You can mark the plastic wrap with a marking pen to label the samples 1, 2, 3 if you wish.

Sample 1
1/8 teaspoon
of baking soda

Sample 2
1/4 teaspoon
of baking soda

Sample 3
1/2 teaspoon
of baking soda

6. Fold up the plastic wrap around the baking soda samples. This will allow you prevent the mixing of the vinegar with the baking soda until you are ready for the reaction to occur.

7. Carefully place "Sample 1" of the wrapped sodium bicarbonate into the bag labeled "1". Try to "float " the wrapped sample on the surface of the vinegar. Carefully seal the bag. Be sure the bag is sealed completely, but do not allow the two reactants to mix yet..

8. From the outside of the sealed Ziploc® bag put your fingers on the package of baking soda. Rub the package with your fingers so the it opens up. Shake the Ziploc® bag so the contents mix. Immediately, some bubbles should form. Watch what happens. Touch the outside of the bag to sense its temperature. Record your observations on the Concept Report Sheet. KEEP this bag sealed and set it aside.

9. Repeat steps 8 and 9 with the other Ziploc® bags.

10. Estimate the amount of gas produced in each bag by carefully rolling the top of the bag down so the gas is trapped in the bottom of the bag. Stop rolling when you meet resistance. The skin of the bag will tighten and look like a balloon. Estimate the relative volumes of the three bags.   (1/4 of a bag full, 1/2 bag full, etc.)

11. Flush the contents of the bags down the drain with a steady flow of water.

Concept Report Sheet
Bridging the Gap   I

Name _____

Sodium Bicarbonate, NaHCO3, and Vinegar, CH3COOH
Heat Effects and Production of CO₂

Observations

Record the heat effects and volume changes you observed when you mixed the sodium bicarbonate with the vinegar.

Mixture	Gas volume observation (A fourth of a bag, etc.)	Temperature observation (No effect, warmer, cooler)
Vinegar and 1/8 teaspoon of baking soda		
Vinegar and 1/4 teaspoon of baking soda		
Vinegar and 1/2 teaspoon of baking soda		

Analysis and Conclusions

1. Is the reaction exothermic or is it endothermic? Justify your answer based on your observations.

2. What happened to the volume of the gas in the Ziploc® bag when you increased the amount of sodium bicarbonate used? Why do you think this happened?

3. Write the overall reaction between vinegar and sodium bicarbonate.
   Circle the reaction product you believe causes the observed volume changes?

4. What do you think would happen if you continued to add larger and larger amounts of baking soda? Do you think the volume would continue to increase or would there be some limit? Why?

Solution to Speaking of Chemistry crossword puzzle

Chemical Reactivity

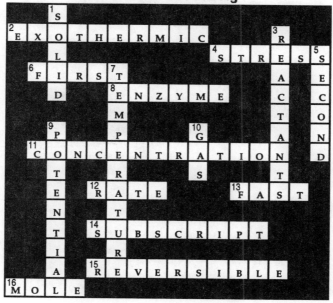

123

9 Acids and Bases --- Chemical Opposites

9.1 Acids

Main topics in this section are the properties of acids, the Arrhenius definition of acids, the behavior of acids in aqueous solutions, acidic oxides, sources of acids, monoprotic acids, polyprotic acids and equations illustrating reactions of acids with water.

Objectives

After studying this section a student should be able to:

describe the characteristic properties of acids
describe the effects of acid on blue litmus
state the Arrhenius definition of an acid
write example equations illustrating the reaction of acids with water
name three common acidic oxides
write example equations illustrating the reaction of acidic oxides with water
write the formula for the hydronium ion
write an equation showing how an acid reacts with an active metal such as calcium, Ca
write the formulas for nitric acid, sulfuric acid and phosphoric acid and give
    a use for each

Key terms

Arrhenius acid	litmus	active metals	hydronium ion
acidic oxides	nonmetal	sulfuric acid	phosphoric acid
nitric acid	acid rain		

9.2 Bases

Main topics in this section are the properties of bases, the Arrhenius definition of bases, the chemical behavior of bases in aqueous solutions, the nature of basic oxides and equations illustrating reactions of bases.

Objectives

After studying this section a student should be able to:

describe the characteristic properties of bases
describe the effect of bases on litmus
state the Arrhenius definition of a base
tell what a basic solution is in terms of relative concentrations of OH^{1-} and H_3O^{1+}
tell what an acidic solution is in terms of relative concentrations of OH^{1-} and H_3O^{1+} give
example equations illustrating the dissolving of a base in water
write example equations illustrating the reaction of basic oxides with water
give examples of basic oxides

Key terms

Arrhenius base	acidic solution	basic solution	hydroxide ion
alkali metal	basic oxides	alkaline earth metals	lime

9.3 Neutralization Reactions

Main topics in this section are the definition of neutralization of Arrhenius acids and bases, salts, role of spectator ions, nature of electrolytes, nonelectrolytes and the concept of the net ionic equation.

Objectives

After studying this section a student should be able to:

describe what is meant by a salt and how they can be formed
describe what neutralization is in terms of Arrhenius acids and bases and give examples
describe what is meant by a spectator ion
tell how a net ionic equation differs from a complete total ionic equation
write example net ionic equations illustrating the neutralization of an acid with base,
　　　　given solublities and strengths of acid and base in the reaction

Key terms

neutralization	salt
electrolyte	nonelectrolyte
net ionic equation	spectator ion
completion of a reaction	equivalent amounts of acid and base
water-soluble salt	ionization

9.4 Brønsted-Lowry Acid-Base Definitions

This section discusses the definition and description of Brønsted-Lowry acids and bases, definitions of proton donor, proton acceptor, conjugate acid, conjugate base, amphiprotic species, and the concept of conjugate acid-base pairs.

Objectives

After studying this section a student should be able to:

state the Brønsted-Lowry definition for an acid
state the Brønsted-Lowry definition for a base
describe neutralization in terms of Brønsted-Lowry acids and bases
write example equations illustrating a Brønsted acid-base reaction
identify the Brønsted-Lowry acid and base in a given reaction
tell how the Brønsted-Lowry theory differs from the Arrhenius theory

Key terms

Brønsted-Lowry acid	Brønsted-Lowry base	conjugate base
proton donor	proton acceptor	conjugate acid
amphiprotic	conjugate acid-base pair	

9.5 The Strengths of Acids and Bases

Main ideas in this section are the descriptions of strong and weak acids and bases, strong electrolytes, weak electrolytes, reason why some hydrogens are acidic and others are not, role of structure on conjugate acid-base relative strengths, behavior of polyprotic acids, and salts of polyprotic acids.

Objectives

After studying this section a student should be able to:

explain what is meant by extent of ionization, 100% ionization and slightly ionized
describe what is meant by a strong acid and a strong base in terms of % ionization
describe what is meant by a weak acid and a weak base in terms of % ionization

126

write equations to illustrate stepwise ionization of polyprotic acids
describe how the relative strength of a polyprotic acid differs for each ionization
write example net ionic equations illustrating the neutralization of an acid with base if
   given solublities and strengths of acid and base in the reaction
tell how many acidic hydrogens are in a polyprotic acid by looking at the formula

Key terms

extent of ionization	weak acid	slightly ionized
strong acid	100% ionized	strong base
spectator ion	weak base	weak electrolyte
strong electrolyte	acidic hydrogens	polar bond
equilibrium	polyprotic acid	second-step ionization
strong acid and weak conjugate base	weak acid and strong conjugate base	

9.6 Concentration of Acid and Base Solutions

This section includes a discussion of the idea of concentration and introduces the concept of molarity as a concentration measure. The calculations involving molarity, grams of solute, molar mass, and volume of solution are explained and illustrated. Dilution problems are explained and examples are shown.

Objectives

After studying this section a student should be able to:
   describe what is meant by concentration
   tell how a strong concentration and a strong acid are differ
   use $M = \dfrac{\#\,moles\ solute}{\#\,liters\ of\ solution}$ to calculate molarity from raw data and to calculate moles of
      solute needed to prepare a solution of specific molarity and volume
   calculate grams of solute needed to prepare a solution of specific molarity and volume
   give ion concentrations for a given strength solution of a strong acid such as 1.5 M $HI_{(aq)}$
   give ion concentrations for a given solution of a strong base such as 2.0 M KOH
   calculate the molarity of a diluted solution given initial molarity, initial and final volume

Key terms

concentration of solution	molar or molarity	M and [ ]
dilution	$M_1V_1 = M_2V_2$	dilute solutions
$M = \dfrac{\#\,moles\ solute}{\#\,liters\ of\ solution}$	# grams solute = # moles x molar mass	

9.7 The pH Scale

Major topics developed in this section are the self ionization of water, the ion product for water, changes in concentrations of H_3O^{1+} and OH^{1-} when acids or bases are added to water, the definition of pH, the relation between pH and acidity, the pH scale range, how to determine pH from H_3O^{1+} and OH^{1-} concentrations, pH indicators, and pH values for common substances.

Objectives

After studying this section a student should be able to:

   describe what is meant by a neutral solution

   state the value for the ion product $[H_3O^{1+}][OH^{1-}]$ for water

   describe how $[H_3O^{1+}]$ and $[OH^{1-}]$ in water are controlled by the constant 1×10^{-14}

   calculate $[H_3O^{1+}]$ from $[OH^{1-}]$ and the ion product for water, $[H_3O^{1+}][OH^{1-}]= 1 \times 10^{-14}$

   calculate $[OH^{1-}]$ from $[H_3O^{1+}]$ and the ion product for water, $[H_3O^{1+}][OH^{1-}]= 1 \times 10^{-14}$

   give the definition for pH and tell the pH ranges for acidic, basic and neutral solutions

   tell whether a solution is acidic , basic or neutral given the pH

   calculate pH for a solution from a known $[H_3O^{1+}]$

   calculate pH for a solution from $[OH^{1-}]$ and the relation $[H_3O^{1+}][OH^{1-}] = 1 \times 10^{-14}$

   determine $[H_3O^{1+}]$ from pH using the relation $[H_3O^{1+}]= 10^{-pH}$

Key terms

neutral	ion product	$[H_3O^{1+}][OH^{1-}]= 1 \times 10^{-14}$
acidic pH range	pH	$[H_3O^{1+}] = 10^{-pH}$
basic pH range	natural indicators	$pH = -\log [H_3O^{1+}]$

9.8   Acid-Base Buffers

This section gives the definition of a buffer, explains how buffers are made and how they work to stabilize pH. The role of acid-base buffers in the human body is introduced. The H_2CO_3 / HCO_3^{1-} and the $H_2PO_4^{1-}$ / HPO_4^{2-} buffer systems are described.

Objectives

After studying this section a student should be able to:

   tell what buffer solutions are

   tell what combinations of compounds are required to make a buffer solution

   tell what is the normal range for blood pH

   name the two conditions that occur when blood pH goes outside the normal range

   describe how a buffer system like H_2CO_3 / HCO_3^{1-} regulates pH when either acid or
        base is added

   write equations showing how a buffer system like H_2CO_3 / HCO_3^{1-} regulates pH when
        either acid or base is added

   write equations showing how a buffer controls pH when hydronium ion is added

   write equations showing how the $H_2PO_4^{1-}$ / HPO_4^{2-} buffer system reacts with added
   hydroxide or hydronium ion to stabilize pH

Key terms

acid-base buffer	blood pH
alkalosis	acidosis
H_2CO_3 / HCO_3^{1-} buffer system	buffer system
$H_2PO_4^{1-}$ / HPO_4^{2-} buffer system	

9.9   Indigestion: Why Reach for an Antacid

The normal pH level of human stomach contents is discussed. Compounds in commercial antacids are described. The effect of antacid tablets on stomach acid is illustrated with reactions.

128

Objectives

After studying this section a student should be able to:

    tell what is normal stomach pH
    tell what acid is secreted by the stomach lining
    describe the connection between "heartburn" and stomach pH
    name two common antacids and their active ingredients.

Key terms

    stomach acid   heartburn   antacid   baking soda

Additional readings

Cramer, Tom. "What's in an antacid?" FDA Consumer v 26 , Jan-Feb 1992: 21.

Davenport, Horace W. "Why The Stomach Does Not Digest Itself." Scientific American Jan. 1972 : 86 .

"Drain Cleaners: Dealing with your drains." Consumer Reports Jan. 1994: 44-48.

Husar, John. " Catch on to the pH factor; better catches may follow." Chicago Tribune 22 October 1989, Sports.

Mebane, R.C.., & Reybold, T. "Edible acid-base indicators." Journal of Chemical Education, 62, (1985): 285.

McNulty, Karen. "Biscuit blues. (acid-base chemistry in the kitchen)." Science World 11 Mar. 1994: 18.

Answers to Odd Numbered Questions for Review and Thought

1.   An Arrhenius acid is a substance that dissolves in water and donates a proton, H^{1+}, to water molecules.

3.   The hydronium ion has the formula H_3O^{1+}, $\left[\begin{array}{c} H:\overset{..}{\underset{..}{O}}:H \\ \overset{..}{H} \end{array}\right]^+$. It is a water molecule with an extra proton bonded to one of the unshared pairs on the oxygen.

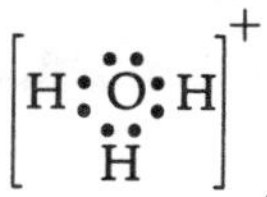

5.   Acidic solutions have H_3O^{1+} concentrations that are greater than the OH^{1-} concentration. The pH is less than 7 for acidic solutions. Basic solutions have OH^{1-} concentrations that are greater than the H_3O^{1+} concentration. Basic solutions have a pH greater than 7. Neutral solutions have pH = 7 and have equal concentrations of hydronium ions, H_3O^{1+}, and hydroxide ions, OH^{1-}.

7.   A salt is the substance formed between the anion of an acid and the cation of a base. An example is KBr, a typical salt formed in the reaction shown here.
$$KOH_{(aq)} + HBr_{(aq)} \longrightarrow KBr_{(aq)} + H_2O_{(\ell)}$$

9.   Spectator ions are essential because they keep electrical "charge balance". They do not participate directly in a reaction or change during the course of a reaction.

11. A Brønsted-Lowry base is any molecule or ion that accepts a proton. In the following equation the water molecule acts as a base and accepts a proton. See section 9.4

$$HCl_{(aq)} \quad + \quad H_2O_{(\ell)} \quad \longrightarrow \quad Cl^{1-}_{(aq)} \quad + H_3O^{1+}_{(aq)}$$

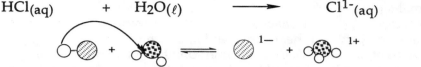

The theory includes nonaqueous solutions; here the ethyl alcohol molecule acts as a base.

$$CH_3CH_2OH + H^{1+} \rightleftharpoons CH_3CH_2OH_2^{1+}$$

13. A proton donor is a molecule or ion that donates an H^{1+} ion such as HI does here

$$HI_{(aq)} + H_2O_{(\ell)} \longrightarrow H_3O^{1+}_{(aq)} + I^{1-}_{(aq)}$$

A proton acceptor is a molecule or ion that bonds to a proton. The $H_2O_{(\ell)}$ above accepts a proton; so does the $NH_{3(aq)}$ in $NH_{3(aq)} + H_3O^{1+}_{(aq)} \rightleftharpoons NH_4^{1+}_{(aq)} + OH^{1-}_{(aq)}$
An amphiprotic substance can act as either a proton donor or a proton acceptor
This example shows water, $H_2O_{(\ell)}$, acting as a base and HBr as the acid,

$$HBr_{(aq)} + H_2O_{(\ell)} \longrightarrow H_3O^{1+}_{(aq)} + Br^{1-}_{(aq)}$$

and this second example shows water acting as an acid with sulfide ion acting as a base.

$$S^{2-}_{(aq)} + H_2O_{(\ell)} \rightleftharpoons HS^{1-}_{(aq)} + OH^{1-}_{(aq)}$$

When a base bonds to a proton, i.e. accepts a proton, the particle that results is called the "conjugate acid" of this base; if the reaction is read backwards, right to left, the conjugate acid acts as the "proton donor" (acid).
A conjugate base is what is left of an acid after it gives up its proton. A conjugate acid-base pair is a set of two particles, one that is a proton donor and the other is the particle formed after the proton is lost. This is illustrated here. See also question 35.

$$\underset{\text{acid}}{H_3PO_{4(aq)}} \rightleftharpoons \underset{\text{conjugate base}}{H_2PO_4^{1-}_{(aq)}} + H^{1+}_{(aq)}$$

15. Common acidic substances are aspirin, coffee, vinegar, fruit juice and carbonated beverages. See section 9.1

17. Generally, nonmetal oxides are acidic and oxides of the metallic representative elements are basic when dissolved in water. See sections 9.1 and 9.2
$SO_{2(g)}$ acidic $CaO_{(s)}$ basic $CO_{2(g)}$ acidic $MgO_{(s)}$ basic $SO_{3(g)}$ acidic

19. a. Vinegar, acidic   b. Citrus fruits, acidic   c. Aspirin, acidic
    d. Lye, basic   e. Black coffee, acidic   f. Milk of Magnesia, basic
    g. Detergents, basic   h. Household ammonia, basic   i. Vitamin C, acidic

21. a. sour taste, acid   b. bitter taste, base
    c. slippery feeling, base   d. change color of red litmus to blue, base
    e. change color of blue litmus to red, acid

23. a. An acidic hydrogen is typically any hydrogen covalently bonded to a highly electronegative atom, usually oxygen or a halogen; it is a hydrogen that can break away easily as H^{1+}, leaving its electron behind.

130

b. A polyprotic acid is an acid with more than one acidic hydrogen; $H_2SO_{4(aq)}$ has two acidic hydrogens attached to two of the oxygens in the molecule.

c. Molarity is a concentration measure equal to the number of moles of solute in one liter of solution; it is calculated by dividing the # moles of solute by the # of liters of solution.

d. Concentration measures the relative amount of dissolved solute in a definite amount of solution.

25. An acid-base indicator changes color with changes in acidity or pH. The indicator molecules act as both acid and base. They display one color when protonated in acidic solutions and another color when they lose a proton in basic solutions. Litmus is blue in basic solutions and turns red in acidic solutions. It is sometimes called "red litmus" when it is red and "blue litmus" when it is blue. See section 9.7

27. Acidosis is a condition resulting from blood pH levels below 7.35, so that oxygen transport is impaired. Alkalosis is a condition that occurs when blood pH rises above 7.5. Human blood usually has a pH around 7.4. See section 9.8

29. $CaO_{(s)} + H_2O_{(\ell)} \longrightarrow Ca(OH)_{2(aq)}$
$Na_2O_{(s)} + H_2O_{(\ell)} \longrightarrow 2\ NaOH_{(aq)}$
$MgO_{(s)} + H_2O_{(\ell)} \longrightarrow Mg(OH)_{2(aq)}$
$Al_2O_{3(s)} + 3\ H_2O_{(\ell)} \longrightarrow 2\ Al(OH)_{3(aq)}$

31. a. The total balanced equation between hydrobromic acid and calcium hydroxide is
$2\ HBr_{(aq)} + Ca(OH)_{2(aq)} \longrightarrow CaBr_{2(aq)} + 2\ H_2O_{(\ell)}$
Strong acid ・・・ Strong base ・・・ Soluble salt ・・・ Nonelectrolyte molecule
Most of the reactants and products are strong electrolytes so the ionic equation is

$2\ H^{1+}_{(aq)} + 2\ Br^{1-}_{(aq)} + Ca^{2+}_{(aq)} + 2\ OH^{1-}_{(aq)} \longrightarrow Ca^{2+}_{(aq)} + 2\ Br^{1-}_{(aq)} + 2\ H_2O_{(\ell)}$

The $Ca^{2+}_{(aq)}$ and $Br^{1-}_{(aq)}$ appear exactly the same on both sides of the equation (they are spectator ions) so they can be deleted to give the net ionic equation. The H^{1+} and OH^{1-} combine to form an H_2O molecule; their bonding changes so the H^{1+}, OH^{1-} and $H_2O_{(\ell)}$ must be kept in the net ionic equation: $H^{1+}_{(aq)} + OH^{1-}_{(aq)} \rightleftharpoons H_2O_{(\ell)}$. This is the same net ionic equation for all strong acid and strong base neutralizations.

b. The balanced total molecular reaction between the strong acid nitric acid and the weak base aluminum hydroxide is shown here.
$3\ HNO_{3(aq)} + Al(OH)_{3(s)} \rightleftharpoons 3\ H_2O_{(\ell)} + Al(NO_3)_{3(aq)}$
The equation showing all ions in solution is
$3\ H^{1+}_{(aq)} + 3\ NO_3^{1-}_{(aq)} + Al(OH)_{3(s)} \rightleftharpoons 3\ H_2O_{(\ell)} + Al^{3+}_{(aq)} + 3\ NO_3^{1-}_{(aq)}$

The HNO_3 is a strong acid and is written as separate ions; $Al(NO_3)_3$ is soluble in water and is written as ions. The $Al(OH)_{3(s)}$ does not ionize readily so it is written as the formula unit. Water is a molecule and a nonelectrolyte so it is shown using the molecular formula. The nitrate ions which appear on both sides can be deleted. The net ionic equation is $3\ H^{1+}_{(aq)} + Al(OH)_{3(s)} \rightleftharpoons 3\ H_2O_{(\ell)} + Al^{3+}_{(aq)}$

33. The stronger acid in a pair of molecules or ions is the one with the greater positive charge.
   a. $H_2O < H_3O^{1+}$ b. $H_2S > HS^{1-}$ c. $H_2SO_4 > HSO_4^{1-}$ d. $H_3PO_4 > H_2PO_4^{1-}$

35. a. $HCl_{(aq)}$ + $H_2O_{(\ell)} \longrightarrow H_3O^{1+}_{(aq)}$ + $Cl^{1-}_{(aq)}$ See section 9.4
   acid   base   conjugate acid   conjugate base

   b. $H_2SO_{4(aq)}$ + $H_2O_{(\ell)} \longrightarrow H_3O^{1+}_{(aq)}$ + $HSO_4^{1-}_{(aq)}$
   acid   base   conjugate acid   conjugate base

   c. $NH_{3(aq)}$ + $H_2O_{(\ell)} \rightleftharpoons OH^{1-}_{(aq)}$ + $NH_4^{1+}_{(aq)}$
   base   acid   conjugate base   conjugate acid

   d. $CH_3COOH_{(aq)}$ + $H_2O_{(\ell)} \rightleftharpoons H_3O^{1+}_{(aq)}$ + $CH_3COO^{1-}_{(aq)}$
   acid   base   conjugate acid   conjugate base

37. A solution with a pH of 2 has an hydronium ion concentration of $10^{-pH} = 10^{-2} = 0.01$, while a solution with pH = 10 has an $H_3O^{1+}_{(aq)}$ ion concentration of $10^{-pH} = 10^{-10}$. The pH 2 solution is much more acidic than the pH 10 solution. See section 9.7.

39. a.   Vinegar, pH = 3.0 is acidic   b.   Baking soda, pH = 8.5 is basic
   c.   Beer, pH = 4.2 is acidic   d.   Lye, pH = 14 is basic
   e.   Soft drinks, pH = 4.5 is acidic   f.   Milk of Magnesia, pH = 10.4 is basic

41. Soap solution has a pH = 10 and gives a $[H_3O^{1+}] = 10^{-pH}$ so $[H_3O^{1+}] = 10^{-10}$. Household ammonia has a pH = 12 and gives a $[H_3O^{1+}] = 10^{-pH}$ so $[H_3O^{1+}] = 10^{-12}$. The relative or comparative acidity of soap to household ammonia can be figured by dividing the soap solution H_3O^{1+} concentration by the household ammonia H_3O^{1+} concentration.

$$\frac{\text{Soap solution} \left[H_3O^+\right]}{\text{Household ammonia} \left[H_3O^+\right]} = \frac{10^{-10} \text{ moles } H_3O^{1+} / \text{liter}}{10^{-12} \text{ moles } H_3O^{1+} / \text{liter}} = 10^{-10 \, - \, (-12)} = 10^{-10+12} = 10^2 = 100$$

Soap solution is 100 times more acidic than household ammonia. See section 9.7.

43. A mixture of $NaHCO_3$ and Na_2CO_3 will act a buffer because the acidic HCO_3^{1-} can provide H^{1+} to neutralize added base. The CO_3^{2-} will react with added acid or H^{1+} to produce HCO_3^{1-}. This action will stabilize the H^{1+} concentration and pH when either acid or base is added. See section 9.8.

45. Baking soda is used to neutralize acid burns because the baking soda is a weak base. It will neutralize the acid, but not create caustic burns from any excess. Baking soda has the formula $NaHCO_3$. The reaction is as follows.
   $NaHCO_{3(aq)}$ + $HCl_{(aq)} \longrightarrow H_2O_{(\ell)}$ + $CO_{2(g)}$ + $NaCl_{(aq)}$ See section 9.3.

Solutions for Selected Problems

1. The number of moles of solute can be figured using molarity $= \dfrac{\text{\# moles solute}}{\text{\# liters of solution}}$.

   The number of moles solute equal the molarity times the volume in liters.
   moles solute = MxV = (molarity) (volume of solution in liters)
   Round answer to the correct number of significant digits, s.d.. Relationships within the metric system are exact, 1 L = 1000 mL , and do not affect the number of s.d. in an answer.

   a. moles HCl $= \dfrac{0.15 \text{ mole HCl}}{1 \text{ liter}} \times \dfrac{200 \text{ mL}}{1} \times \dfrac{1 \text{ liter}}{1000 \text{ mL}} = 0.030$ mole HCl

   c. moles $HNO_3 = \dfrac{0.25 \text{ mol } HNO_3}{1 \text{ liter}} \times \dfrac{0.050 \text{ liter}}{1} = 0.0125$ moles HNO_3.

   Round off to 0.012 moles, with two significant digits, because 0.25 and 0.50 have 2 s.d.

2. The number of moles of solute in a milliliter of solution is calculated using the relation that 1 mL of solution equals 0.001 Liter.

   moles solute = MxV = (molarity)(volume of solution in liters)

   The number of moles in one milliliter is equal to the original molarity multiplied by 0.001 which moves the decimal point three places to the left

   a. moles HCl $= \dfrac{6.0 \text{ mole HCl}}{1 \text{ liter}} \times \dfrac{1 \text{ mL}}{1} \times \dfrac{1 \text{ L}}{1000 \text{ mL}} = \dfrac{6.0 \text{ mole HCl}}{1 \text{ liter}} \times \dfrac{0.001 \text{ L}}{1}$
   moles HCl = 0.0060 moles HCl

   b. moles KOH $= \dfrac{2.5 \times 10^{-2} \text{ mole KOH}}{1 \text{ liter}} \times \dfrac{0.001 \text{ L}}{1} = 0.000025$ moles KOH

3. The number of grams of solute in a solution can be calculated using the molarity and the volume in liters as well as the molar mass of the solute. The number of moles is determined from the molarity and the volume of the solution. See problem 1.
   moles solute = MxV = (molarity) (volume of solution in liters)
   Then, g solute is calculated from g solute = moles x molar mass of solute.
   An alternative way to solve this problem is to first determine the moles of solute in one mL as was done in problem 2. Calculate the total moles of solute by multiplying the moles in one mL by the number of mLs. The number of grams of solute can be calculated by multiplying number of moles by the molar mass.

   a. Determine the molar mass for the solute by adding the mass each element contributes.
   Molar mass for NaOH = 22.99 g Na + 16.00 g O + 1.01g H = 40.00 g NaOH
   Convert the volume of solution to liters (53 mL/ 1)(1L / 1000 mL) = 0.053 L
   Use the molarity and volume to determine the number of moles of solute.
   moles NaOH = MxV = (1.5 mole NaOH / L) x (0.053 L) = 0.0795 moles NaOH
   Calculate the number of grams of solute using the moles of solute and the molar mass:
   g NaOH = moles solute x molar mass

   g NaOH = 0.0795 molesNaOH $\times \dfrac{40.00 \text{g NaOH}}{1 \text{ mole NaOH}} = 3.18$ g NaOH
   The calculation can be done in one set up:

$$\text{g NaOH} = \frac{1.5 \text{ mole NaOH}}{1 \text{ liter}} \times \frac{0.053 \text{ L}}{1} \times \frac{40.0 \text{ g NaOH}}{1 \text{ mole}} = 3.18 \text{ g NaOH} = 3.2 \text{ g NaOH}$$

The 3.18 g NaOH is rounded off to 3.2 g because the 1.5 M NaOH and the 0.053 L have two significant digits each.

b. Molar mass NH_3 = 1(14.00 g N) + 3(1.01g H) = 17.03 g NH_3 = 17.03 g NH_3

$$\text{g NH}_3 = \frac{5.0 \text{ mole NH}_3}{1 \text{ liter}} \times \frac{0.370 \text{ L}}{1} \times \frac{17.03 \text{ g NH}_3}{1 \text{ mole NH}_3} = 31.51 \text{ g NH}_3 = 32 \text{ g NH}_3$$

The 31.51 g NH_3 should be rounded off to 32 g NH_3 two significant digits.

c. Molar mass H_2SO_4 = 2x1.01 g H + 1x32.06 g S + 4x16.00 g O = 98.08 g H_2SO_4

$$\text{g H}_2\text{SO}_4 = \frac{0.35 \text{ mole H}_2\text{SO}_4}{1 \text{ liter}} \times \frac{85 \text{ mL}}{1} \times \frac{1 \text{ L}}{1000 \text{ mL}} \times \frac{98.08 \text{ g H}_2\text{SO}_4}{1 \text{ mole H}_2\text{SO}_4} = 2.92 \text{ g H}_2\text{SO}_4$$

The answer rounded off to two significant digits is 2.9 g H_2SO_4 .

d. Molar mass H_3PO_4 = 3x1.01 g H + 1x30.97 g P + 4x16.00 g O = 98.00 g H_3PO_4.

$$\text{g H}_3\text{PO}_4 = \frac{3.0 \text{ mole H}_3\text{PO}_4}{1 \text{ liter}} \times \frac{0.051 \text{ L}}{1} \times \frac{98.00 \text{ g H}_3\text{PO}_4}{1 \text{ mole H}_3\text{PO}_4} = 14.99 \text{ g H}_3\text{PO}_4$$

The mass of solute should be rounded off to 15 g H_3PO_4 .

5. a. Molar mass CH_3COOH =
$$= \text{2x12.01 g C + 4x1.01 g H + 2x16.00 g O = 60.06 g CH}_3\text{COOH}$$

$$\text{g CH}_3\text{COOH} = 1.0 \text{ liter} \times \frac{0.25 \text{ mole CH}_3\text{COOH}}{1 \text{ liter}} \times \frac{60.06 \text{ g CH}_3\text{COOH}}{1 \text{ mole CH}_3\text{COOH}} = 15. \text{ g CH}_3\text{COOH}$$

b. Molar mass $Al(OH)_3$ = 26.98 g Al + 3x1.01 g H + 3x16.00 g O = 78.01 g $Al(OH)_3$

$$\text{g Al(OH)}_3 = 1.0 \text{ liter} \times \frac{4.0 \text{ mole Al(OH)}_3}{1 \text{ liter}} \times \frac{78.01 \text{ g Al(OH)}_3}{1 \text{ mole Al(OH)}_3} = 312 \text{ g Al(OH)}_3$$

= 310 g $Al(OH)_3$/ liter   Round off answer to two significant digits.

c. g HCl = $1.0 \text{ liter} \times \dfrac{2.5 \times 10^{-3} \text{ mole HCl}}{1 \text{ liter}} \times \dfrac{36.46 \text{ g HCl}}{1 \text{ mole HCl}} = 0.0912 \text{ g HCl} = 0.091 \text{g HCl}$

d. g KOH = $1.0 \text{ liter} \times \dfrac{9.0 \text{ mole KOH}}{1 \text{ liter}} \times \dfrac{56.02 \text{ g KOH}}{1 \text{ mole KOH}} = 504 \text{ g KOH} = 500 \text{ g KOH}$

Answer needs to have two significant digits and can be written 5.0×10^2 to show this.

6. When a solution is diluted by the addition of more solvent the number of moles of solute at the beginning equals the number of moles of solute at the end. The formal expression is $M_1V_1 = M_2V_2$ where the subscript 2 indicates the final condition after dilution and subscript 1 indicates the initial condition. One way to solve these problems is to identify the numerical value for each quantity in this expression; substitute into the equation and solve for the unknown. Both volumes must have the same units, usually liters.

a. Use the equation $M_1V_1 = M_2V_2$;

    identify M_1 = 2.5 MHCl ; V_1 = 180 mL = 0.180 L
           M_2 = unknown ; V_2 = 0.50 L
    substitute values (2.5 MHCl)(0.180 L) = (M_2)(0.50 L); solve for M_2.

    M_2 = 2.5 M HCl $\times \dfrac{0.180 \text{ L}}{0.50 \text{ L}}$ = 0.90 M HCl

b. Identify M_1 = 6.0 M NH_3 ; V_1 = 30 mL = 0.030 L
           M_2 = unknown ; V_2 = 1000mL = 1.0 L

    M_2 = 6.0 M NH_3 $\times \dfrac{0.030 \text{ L}}{1.0 \text{ L}}$ or M_2= 6.0 M NH_3 $\times \dfrac{30 \text{ mL}}{1000 \text{ mL}}$ = 0.18 M NH_3

c. Identify M_1 = 12.0 M HCl; V_1 = 1200 mL = 1.2 L
           M_2 = unknown ; V_2 = 10000 mL = 10.0 L

    M_2 = 12.0 M HCl $\times \dfrac{1.2 \text{ L}}{10.0 \text{ L}}$ or M_2 = 12.0 M HCl $\times \dfrac{1200 \text{ mL}}{10000 \text{ mL}}$ =1.44 M HCl =1.4 M HCl

d. Identify M_1 = 0.2 M Al(OH)$_3$; V_1 = 48 mL = 0.048L
           M_2 = unknown ; V_2 = 500 mL = 0.500 L

    M_2 = 0.2 M Al(OH)$_3 \times \dfrac{0.048 \text{ L}}{0.50 \text{ L}}$ or M_2 = 0.2 M Al(OH)$_3 \times \dfrac{48 \text{ mL}}{500 \text{ mL}}$ = 0.0192M Al(OH)$_3$
    The answer rounded to 1 significant digit is 0.02 M Al(OH)$_3$.

7. These dilution problems are solved in the same fashion as number 6 above.

    a. (3.0 M HBr)(75 mL) = (M$_2$)(500 mL); M_2 = 3.0 M HBr $\times \dfrac{75 \text{ mL}}{500 \text{ mL}}$ = 0.45 M HBr

    b. (18 M H_2SO_4)(10 mL) = (M$_2$)(1000 mL);

    M_2= 18 M $H_2SO_4 \times \dfrac{10 \text{ mL}}{1000 \text{ mL}}$= 0.18 M H_2SO_4

    c. (8.0 M HNO_3)(700 mL) = (M$_2$)(2000 mL); M_2 = 8.0 M $HNO_3 \times \dfrac{700 \text{ mL}}{2000 \text{ mL}}$ or

    M_2 = 8.0 M $HNO_3 \times \dfrac{0.700 \text{ L}}{2.0 \text{ L}}$ =2 .8 M HNO_3

    d. (0.2 M NaOH)(65 mL) = (M$_2$)(100 mL); M_2 = 0.2 M NaOH$\times \dfrac{65 \text{ mL}}{100 \text{ mL}}$ or

    M_2 = 0.2 M NaOH$\times \dfrac{0.065 \text{ L}}{0.100 \text{ L}}$ = 0.13 M NaOH = 0.1 M NaOH   The answer is
    rounded to one significant digit because there is only one s.d. in 0.2 M NaOH.

9. The molarity is determined from the volume of solution in liters and the number of moles of solute. The solute must be converted to moles if it is in grams. The molar mass is needed to convert grams to moles. The volume of solution must be converted to liters if it is in any other volume unit. The answer is rounded to the correct number of significant digits.

a. The 65.0 g NaOH must be converted to moles of NaOH. The molar mass is 40.00 g NaOH. (See problem 3a). The 865 mL of solution must be converted to liters.

$$\text{Molarity NaOH} = 65.0 \text{ g NaOH} \times \frac{1 \text{ mole NaOH}}{40.00 \text{ g NaOH}} \times \frac{1}{0.865 \text{ L}} = 1.88 \text{ M NaOH}$$

b. $$\text{Molarity H}_2\text{SO}_4 = 1.5 \text{ g H}_2\text{SO}_4 \times \frac{1 \text{ mole H}_2\text{SO}_4}{98.08 \text{ g H}_2\text{SO}_4} \times \frac{1}{0.035 \text{ L}} = \; = 0.44 \text{ M H}_2\text{SO}_4$$

c. $$\text{Molarity Ba(OH)}_2 = 1.5 \text{ g Ba(OH)}_2 \times \frac{1 \text{ mole Ba(OH)}_2}{171.35 \text{ g Ba(OH)}_2} \times \frac{1}{0.475 \text{ L}} = 0.018 \text{ M Ba(OH)}_2$$

d. $$\text{Molarity Ca(OH)}_2 = 12.6 \text{ g Ca(OH)}_2 \times \frac{1 \text{ mole Ca(OH)}_2}{74.10 \text{ g Ca(OH)}_2} \times \frac{1}{18 \text{ L}} = 0.0094 \text{ M Ca(OH)}_2$$

11. a. $$M_2 = 6.0 \text{ M NH}_3 \times \frac{0.055 \text{ L}}{2.0 \text{ L}} = 0.165 \text{ M NH}_3 = 0.16 \text{ M NH}_3$$

b. $$M_2 = 1.5 \text{ M HCl} \times \frac{0.400 \text{ L}}{2.0 \text{ L}} = 0.30 \text{ M HCl}$$

c. $$M_2 = 18. \text{ M H}_2\text{SO}_4 \times \frac{0.628 \text{ L}}{2.0 \text{ L}} = 5.652 \text{ M H}_2\text{SO}_4 = 5.7 \text{ M H}_2\text{SO}_4$$

d. $$M_2 = 0.50 \text{ M KOH} \times \frac{0.125 \text{ L}}{2.0 \text{ L}} = 0.03125 \text{ M KOH} = 0.031 \text{ M KOH}$$

12. The pH for a solution is determined from the definition $pH = -\log[H_3O^{1+}]$. A simple graphic relationship between the hydronium ion concentration and pH is shown here. The top row gives the concentration $[H_3O^{1+}]$ and the second line gives the matching pH . This is limited only to the whole numbered exponents of 10.

$[H_3O^{1+}]$	10^0	10^{-1}	10^{-2}	10^{-3}	10^{-4}	10^{-5}	10^{-6}	10^{-7}	10^{-8}	10^{-9}	10^{-10}	10^{-11}	10^{-12}	10^{-13}	10^{-14}
pH	0	1	2	3	4	5	6	7	8	9	10	11	12	13	14

a. Hydrochloric acid, HCl(aq) is a strong acid. The $[H_3O^{1+}] = 1.0 \times 10^{-2}$ because strong acids are 100% ionized and the ion concentrations equal the original acid concentration. $pH = -\log[H_3O^{1+}]$; $pH = -\log 1 \times 10^{-2}$;
The log function on a calculator gives pH = 2
or the graphical relationship shown above can be used to get the same result..

b. Nitric acid, HNO$_3$(aq), is a strong acid, so the $[H_3O^{1+}] = [NO_3^{1-}] = 1.0 \times 10^{-3}$
$pH = -\log[H_3O^{1+}]$; $pH = -\log 1 \times 10^{-3}$; $pH = 3$

136

c. Sodium hydroxide is a strong base. The Na^{1+} and OH^{1-} concentrations equal the original NaOH concentration. $[Na^+] = [OH^{1-}] = 0.1 = 1 \times 10^{-1}$. Use the relationship $Kw = [H_3O^{1+}][OH^{1-}] = 1 \times 10^{-14}$ to calculate the $[H_3O^{1+}]$. Substitute 1×10^{-1} for the hydroxide concentration; $1 \times 10^{-14} = [H_3O^{1+}][OH^{1-}] = [H_3O^{1+}][1 \times 10^{-1}]$;

$[H_3O^{1+}] = \dfrac{1 \times 10^{-14}}{[1 \times 10^{-1}]} = \dfrac{1}{1} \times 10^{-14--1} = 1 \times 10^{-14+1} = 1 \times 10^{-13}$. When the coefficients are 1, a simpler method recognizes that the sum of the exponents on ten on both sides must equal -14 because $[H_3O^{1+}][OH^{1-}] = 1 \times 10^{-14}$. This means $-1 + ? = -14$; The unknown exponent must be -13. so $[H_3O^{1+}] = 1 \times 10^{-13}$.
$pH = -\log[H_3O^{1+}]$; $pH = -\log 1 \times 10^{-13}$; $pH = 13$.

d. Hydrobromic acid, HBr, ionizes 100%. The concentration of hydronium ion, $[H_3O^{1+}]$, is 0.10 M = 1.0×10^{-1}.
Using $pH = -\log[H_3O^{1+}]$; $pH = -\log 1 \times 10^{-1}$; $pH = 1$

13. d. Neutral water has equal concentrations for $[H_3O^{1+}]$ and $[OH^{1-}]$.
$Kw = [H_3O^{1+}][OH^{1-}] = 1 \times 10^{-14}$; Let x = $[H_3O^{1+}] = [OH^{1-}]$

$[x][x] = 1 \times 10^{-14}$; $x^2 = 1 \times 10^{-14}$; $x = \sqrt{1 \times 10^{-14}} = \sqrt{1} \times 10^{-14+2} = 1 \times 10^{-7}$;
$pH = -\log[H_3O^{1+}]$; $pH = -\log 1 \times 10^{-7}$; $pH = 7$

14. a. pH = 1 ; $[H_3O^{1+}] = 10^{-pH}$; $[H_3O^{1+}] = 10^{-1} = 1 \times 10^{-1}$ or 0.1
    b. pH = 0 ; $[H_3O^{1+}] = 10^{-pH}$; $[H_3O^{1+}] = 10^{-0} = 1 \times 10^{-0}$ or 1

15. a. pH = 8 ; $[H_3O^{1+}] = 10^{-pH}$; $[H_3O^{1+}] = 10^{-8} = 1 \times 10^{-8}$;
    $Kw = [H_3O^{1+}][OH^{1-}] = 1 \times 10^{-14}$; substitute for $[H_3O^{1+}]$.
    $1 \times 10^{-8} [OH^{1-}] = 1 \times 10^{-14}$;
    When the multipliers are 1, a simpler method recognizes that the sum of the exponents on ten on both sides must equal -14.
    This means $-8 + ? = -14$;
    The unknown exponent must be -6. so $[OH^{1-}] = 1 \times 10^{-6}$
    A more formal method is to solve for $[OH^{1-}]$.
    $[OH^{1-}] = (1 \times 10^{-14}) \div 1 \times 10^{-8} = (1 \times 10^{-14})(1 \times 10^8) = 1 \times 10^{-14+8} = 1 \times 10^{-6}$
    $[OH^{1-}] = 1 \times 10^{-6}$ The results are the same.

    b. pH = 10; $[H_3O^{1+}] = 10^{-pH}$; $[H_3O^{1+}] = 10^{-10} = 1 \times 10^{-10}$
    $Kw = [H_3O^{1+}][OH^{1-}] = 1 \times 10^{-14}$; $[1 \times 10^{-10}][OH^{1-}] = 1 \times 10^{-14}$
    $[OH^{1-}] = 1 \times 10^{-4}$

Speaking of Chemistry

Name _____

Acids and Bases
Chemical Opposites

Across

1 Describes substances that do not contain water
3 Compounds like CaO and CO_2 that contain oxygen
7 Substance that neutralizes acids
8 A weak base with formula NH_3
10 A basic metal oxide with formula, CaO
11 Solutions with pH below 7 are ____
14 The reaction between an acid and a base
17 Acids have a tart or _____ taste
18 The pH of a solution with $H^+ = 10^{-5}$
19 An antacid that contains $CaCO_3$
20 Ions that are in a solution but do not particitate in reactions

Down

1 Substance that neutralizes stomach acid
2 Common cleaning product that is a base
3 The pH for a 1 M HCl solution
4 An acid like HCl is a proton _____
5 Basic solutions are _____ to the touch.
6 A dye that changes color when pH changes
9 Compounds with formulas that include separate water
molecules, $CaSO_4 \cdot 2\ H_2O$
12 Acids are defined as _____ donors
13 Describes acids and bases that do not ionize completely
15 A solution with excess OH^- is basic or

_____
16 Particle with an charge
19 The number of acidic protons in H_2SO_4

140

Bridging the Gap   I
Acidity, pH, Indicator Paper and Red Cabbage

Acids and bases are part of us and our surroundings. The body uses hydrochloric acid, HCl(aq) in the stomach during digestion. Calcium carbonate, $CaCO_3$(s), is the active ingredient in Tums™ an antacid used to treat excess stomach acid. These substances all yield solutions with characteristic pH values. Several pH sensitive dyes can be used to estimate pH values for solutions. You will use these as your pH indicator.

Red cabbage contains natural dyes that change color with acidity changes. The color displayed by these dyes is a good indicator of the pH. The color of these dyes is shown in the text in Figure 9.6 on page 267.

This plot is a graphical summary of the relationship between color and pH. Acid solutions will have a red or pink color. Neutral solutions will have a pretty lavender color and basic solutions will have a blue-green or yellow color.

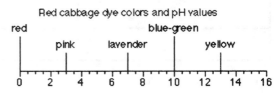

Red cabbage dye colors and pH values

Red cabbage indicator solution

Prepare the red cabbage dye indicator solution using the procedure on page 268 of the text. Check results against the listing of pH values in Figure 9.5 on page 266.
1. Test at least three different materials; each must be dissolved in water. Place each solution in a clear, colorless container such as a glass or jelly jar.
2. Place a piece of white paper on your work surface. This gives a good background for seeing the solution colors. Label the different containers.
3. Add about 10 mL, 2 teaspoons, of cabbage extract to each solution and stir to do a test. Record the results of the different tests in the table on the Concept Report Sheet.

Indicator paper
1. Take a 2-inch-wide strip of white facial tissue and tape the upper end to a pencil so the tissue hangs down and can be lowered into the cabbage indicator solution. Put about 1/3 cup of cabbage dye into a quart jar or other tall container. Lower the strip of tissue paper into the jar until its lower end is in the cabbage dye. Rest the pencil across the top of the container so the tissue hangs inside. Let the solution wick up the paper until the liquid has climbed to about 6 inches. This will take about 40 minutes.
2. Remove the wet paper from the liquid and hang it up to dry.
3. When the tissue is thoroughly dry cut it into pieces about 1 inch x 1 inch. Store them in a brown jar or dark place.
4. This paper will act as an acid base indicator. It will turn pink when touched with a drop of acid and turn blue-green when touched with a drop of base; try it. Save the indicator paper for your own tests.

Concept Report Sheet
Bridging the Gap   I
Acidity, pH, Indicator Paper and Red Cabbage

Name_____

Data and Observations

Household substance	Cabbage juice indicator color	Estimated pH	pH using page 266

Analysis and Conclusions

Do you believe the estimated experimental pH values and the table values agree? Remember these are only estimates and are not exact. Justify your answer.

Bridging the Gap   II
Acid Concentration and Red Cabbage Dyes

A common base is household ammonia which is a mixture of water and ammonia, $NH_{3}(g)$. The strong, biting odor is from the dissolved $NH_{3}(g)$ that escapes from the solution. Household ammonia that has been open for a time will not be as strong as a "fresh" solution. Vinegar is an acid solution which contains acetic acid, $HC_2H_3O_2$, and water. In this exercise you will determine the relative strength of your vinegar and compare the result with the concentration listed in the text. Natural red cabbage dye indicator is used to tell when the solution pH is neutral and the ammonia has just neutralized the vinegar.

Equipment and materials

Five clear colorless glass or acrylic tumblers that can hold 1 cup of liquid, plastic straws, white paper, clear uncolored household ammonia, white vinegar, distilled water or tap water.
1.   Place the piece of white paper on your work surface. This gives a good background for seeing the solution colors. Mark the positions for the different containers.
2.   Place one glass in the middle. Place a glass on either side of the first one.
3.   Pour approximately 30 mL (2 tablespoons) of vinegar into the glass on the left. Rinse your tablespoon with tap water. Add a similar amount of household ammonia to the glass on the right. Place a plastic straw in each glass.
4.   Add two teaspoons of cabbage juice to the middle glass. Add 1/2 cup of water to this glass of cabbage juice. Check to see what the color is for this water mixture.
5.   Add approximately 30 mL of household ammonia and 2 teaspoons of cabbage dye to the 4th glass; mix and set aside to the far right as a reference for the base color.
6.   Place approximately 30 mL of vinegar and 2 teaspoons of cabbage dye in the 5th glass; mix and set aside to the far left as a reference for the acid color.
7.   Use the straw to transfer <u>exactly 20 drops</u> of plain vinegar to the middle glass that already contains the water and cabbage juice mixture. Do not draw any liquid into the straw using your mouth. The straw will collect some liquid inside because it is open and there will be some capillary rise. Cap off the top of the straw with a finger. Some liquid will stay in the straw.

Carefully move the straw over to the middle glass and squeeze the straw to deliver individual drops of vinegar. You will have to do a series of transfers to get 20 drops. Note the color.

8.   Now transfer ammonia solution <u>one drop at a time</u> to the middle glass containing the mixture of water, cabbage dye and 20 drops of vinegar. Swirl or stir and note the color after each drop. Count the number of drops of ammonia needed to produce the lavender color of a neutral solution.
9.   Record the number of drops of ammonia solution used to neutralize 20 drops of vinegar. Based on your work, which of the two solutions has the higher concentration? Use the ratio of # drops of ammonia to # drops of vinegar to find the molarity of the vinegar.
10.   Wash out the middle glass, rinse with tap water; repeat Steps 3, 7, 8, and 9.

143

Bridging the Gap   II   Name _____
Acid Concentration using Red Cabbage Dyes

Data and Observations

	Trial 1	Trial 2
Number of drops of vinegar	20 drops	20 drops
Number of drops of ammonia		
Ammonia is 2.5 M NH_3 (Text, pages 271-272) Calculate the molarity of your vinegar as follows: $M_{vinegar} = 2.5 \times \dfrac{\text{\# drops ammonia}}{\text{\# drops vinegar}} =$		

Analysis and Conclusions

1.   What is the average molarity of the vinegar based on your measurements?

2.   What is the molarity for acetic acid in vinegar as listed in your text?

3.   Do you feel your experimental value agrees with the concentration listed in the text?

4.   Name another acidic solution you could analyze with this neutralization method.

Solution to Speaking of Chemistry crossword puzzle

Acids and Bases
Chemical Opposites

¹A	N	H	Y	D	R	O	U	²S		³O	X	I	⁴D	E	⁵S
N								O		N			O		L
T				⁶I		⁷B	A	S	E				N		I
⁸A	M	M	O	N	I	A		P					O		P
C				D				⁹H					R		P
I			¹⁰L	I	M	E		Y				¹²P			E
D				C		¹¹A	C	I	D			R			R
	¹³W			A				R				R			Y
¹⁴N	E	U	T	R	¹⁵A	L	¹⁶I	Z	A	T	I	O	N		
	A			O		L		O				T			
	K			R		K		N	E	¹⁷S	O	U	R		
						A		N				N			
						L									
				¹⁸F	I	V	E		¹⁹T	U	M	S			
				N				W							
				²⁰S	P	E	C	T	A	T	O	R			

10   Oxidation-Reduction in Chemistry

10.1   Oxygen--The Element

This section describes the abundance of oxygen and the variety of oxygen-containing compounds. It tabulates physical properties for oxygen, O_2, and gives a list of common oxides.

Objectives

After studying this section a student should be able to:

> state the approximate percentage of O_2 in the atmosphere
> describe selected physical properties of oxygen
> give the definition for an oxide and give formulas and names for common oxides
> name three elements that do not form oxides readily

Key terms

     oxygen   oxidation   oxide

10.2   So What is Oxidation, Anyway?

This section gives three different ways to identify an oxidation reaction: gain of oxygen, loss of hydrogen, loss of electrons. Example reactions are used to illustrate each definition. The purpose of automobile catalytic converters is explained. Combustion is defined. Oxidizing agents are described.

Objectives

After studying this section a student should be able to:

> state three definitions for oxidation and give an example of each
> identify the oxidized substance in a reaction
> inspect a reaction equation and tell whether or not oxidation occurred
> give the definition for an oxidizing agent and give applications for oxidizing agents like F_2
> tell what an automobile catalytic converter does to CO
> give a definition and example for combustion

Key terms

oxidation	oxidation products	oxidized
combustion	catalytic converter	hydrogen loss
loss of electrons	disinfectant, bleach	oxidizing agent

10.3   Hydrogen--The Element

This section describes the abundance and sources of H_2. It tabulates some physical properties of hydrogen and lists some common hydrides such as CH_4, NH_3, LiH. Use of hydrogen for lighter-than-air ships (Hindenburg) and as a fuel (space shuttle) are described.

Objectives

After studying this section a student should be able to:

> name classes of compounds that contain hydrogen
> describe selected physical properties of H_2 such as density, color
> name and give formulas for some common hydrides
> describe two commercial applications for H_2 and tell why its use poses a risk

Key terms

hydride
decomposition of water

hydrocarbons
catalytic decomposition

flammable
dehydrogenation

10.4 Reduction--The Opposite of Oxidation

This section gives three ways to identify a reduction reaction: loss of oxygen, gain of hydrogen, gain of electrons. Example reactions are used to illustrate each definition. Reducing agents are defined.

Objectives

After studying this section a student should be able to:

give the definition for a hydride
state the three definitions for reduction
illustrate the three definitions for reduction with reactions
give a definition for a reducing agent
write the reaction for making ammonia by the Haber process and explain its importance
inspect a reaction and identify the reducing agent
inspect a reaction and tell whether or not reduction occurred
give some examples and uses of oxidizing agents
give some examples and uses of reducing agents

Key terms

reduction
reducing agent

gain of hydrogen
loss of oxygen

oxidation-reduction reaction
gain of valence electrons

10.5 The Strengths of Oxidizing and Reducing Agents

This section describes variations in oxidizing and reducing abilities for elements. The concept of activity is introduced and an activity series is listed and applied. Active metals are defined; definitions for strong and weak reducing agents are explained. The concept of a favorable reaction is described.

Objectives

After studying this section a student should be able to:

give a definition for activity and activity series
state definitions for oxidizing agent and reducing agent
give the definition for active metals
use an activity series table to tell the direction for an oxidation-reduction reaction
use an activity series table to identify the stronger oxidizing agents
use an activity series table to predict whether of not a reaction will occur between two reactants

Key terms

activity
weakest oxidizing agent
active metals

activity series
strongest reducing agent

strongest oxidizing agent
weakest reducing agent

10.6 Corrosion--Unwanted Oxidation-Reduction

This section defines corrosion. It gives the reactions for corrosion of iron and tells what requirements exist for rust formation. Galvanizing is described. Other methods for protecting iron are listed.

148

Objectives

After studying this section a student should be able to:

give a definition for corrosion

give a formula for rust and tell why it is a problem

tell what reactants are required for rust to form

explain how rusting can be prevented

describe what galvanizing is and how it prevents rusting

Key terms

corrosion	rust	salt bridge
electrochemical cell	protective coating	galvanizing

10.7   Batteries

Batteries and electrochemical cells are defined. The components of a battery are described. Primary batteries, nonrechargeable ones, are defined and illustrated. Reusable, secondary, batteries are described. Reactions and applications are given for lead-storage, Ni-Cad, and lithium-ion batteries.

Objectives

After studying this section a student should be able to:

define electrochemical cell, anode, and cathode

sketch and label a diagram for an electrochemical cell (see page 298)

describe the functions of anode, cathode, salt bridge

define primary battery and secondary battery give an example of each

classify the following batteries as either primary or secondary: lead-storage, Ni-Cad,
    mercury , lithium-ion, Leclanché dry cell, alkaline dry cell

explain what memory means for a Ni-Cad battery

describe the hazards posed by lead-storage and by mercury batteries

Key terms

electrochemical cell	battery	two electrodes
anode	cathode	salt bridge
electric current	throw-away battery	primary battery
dry cell	renewable	alkaline battery
lithium battery	secondary battery	lead-acid battery
mercury battery	recharging	nickel-cadmium battery
Ni-Cad	memory	toxic ingredients
lithium-ion battery	high energy output	

10.8   Fuel Cells

Fuel cells are defined and applications for fuel cells are given. The hydrogen-oxygen fuel cell is diagrammed and described. Applications of fuel cells in electric automobiles and the space shuttle are discussed.

Objectives

After studying this section a student should be able to:

state the definition for a fuel cell and tell how they differ from batteries

write the reaction that occurs in the hydrogen-oxygen fuel cell

label a diagram of the hydrogen-oxygen fuel cell

Key terms

fuel cell	alkaline fuel cells	hydrogen-oxygen fuel cell

10.9 Chemical Reactions Caused by Electron Flow

This section defines electrolysis and explains how electrical energy is used to produce chemical change. Electroplating of copper and other metals is described. The electrolysis of water to produce hydrogen in summarized. Important chemicals produced by electrolysis are listed.

Objectives

After studying this section a student should be able to:

  give a definition for electrolysis
  label a diagram of a copper electroplating cell and describe the electroplating process
  write the net reaction for the decomposition of water by electrolysis
  write the reduction reaction for electrolysis of water
  write the oxidation reaction for electrolysis of water
  name four important chemicals prepared using electrolysis

Key terms

electrolysis reaction	electron flow	evolved
deposited	electroplating	plating

Additional Readings

Berendsohn, Roy & Rosario Copotosto. "Rust Prevention" <u>Popular Mechanics</u> Sept. 1993: 59.

Bergens, Steven H. et al. "A Redox Fuel Cell That Operates With Methane As Fuel At 120 °C" <u>Science</u> Sept. 2, 1994: 1418.

Consumer Reports Staff. "Best Bets in Auto Batteries" <u>Consumer Reports</u> Oct. 1995: 674.

Consumer Reports Staff. "A 'Greener' Rechargeable Battery?" <u>Consumer Reports</u> Aug. 1993: 489.

Lave, Lester B. & Chris T. Hendrickson, Francis Clay McMichael. "Environmental Implications of Electric Cars" <u>Science</u> May 19, 1995: 993.

Lecard, Marc. "Recharge It" <u>Sierra</u> Nov.-Dec. 1993: 42.

Lipkin, Richard. "Firing Up Fuel Cells: Has a Space-age Technology Finally Come of Age for Civilians?" <u>Science News</u> Nov. 13, 1993: 314.

Moskal, Brian S. "It's the Battery, Stupid!" <u>Industry Week</u> Oct. 3, 1994: 22.

Vizard, Frank. "Strips of Power" <u>Popular Mechanics</u> July, 1994: 98.

Answers to Odd Numbered Questions for Review and Thought

1. a. Oxidation is the loss of electrons such as in $Fe \longrightarrow Fe^{2+}(aq) + 2\ e^-$
    b. Reduction is the gain of electrons as in $F_2(g) + 2\ e^- \longrightarrow 2\ F^-$
    c. Oxidation is the gain of oxygen such as $2\ Ca(s) + O_2(g) \longrightarrow 2CaO(s)$
    d. Reduction is the gain of hydrogen as in $CH_2CH_2(g) + H_2(g) \longrightarrow CH_3CH_3(g)$
    e. Oxidizing agent is a substance that accepts electrons such as the Fe^{3+} in
        $Fe^{3+}(aq) + e^- \longrightarrow Fe^{2+}(aq)$
    f. Reducing agent is a substance that gives up electrons such as the $Zn(s)$ in
        $Zn(s) \longrightarrow Zn^{2+}(aq) + 2\ e^-$

3. a. The carbon has more oxygens in CO_2 than in CO. Remember the O usually has an oxidation number of -2 in compounds The oxidation number for carbon is +4 in CO_2 since one C goes with two O. Carbon has an oxidation number of only +2 in CO since one C goes with one O.

    b. The nitrogen has more oxygens in NO_2 than in NO. The oxidation number is +4 in NO_2 and only +2 in NO.

    c. The sulfur is more oxidized in SO_3 than in SO_2. The combined oxidation number for the three O's is $-6 = -2 \times 3$ so the sulfur is +6 in SO_3. In the SO_2 the two O's have a combined oxidation number of $-4 = -2 \times 2$ so the sulfur is + 4 in for SO_2 .

    d. The chromium is more oxidized in CrO_3 than in CrO because it has more oxygens and has an oxidation number of +6 rather than +2.

    e. The calcium has an oxidation number of +2 in CaO and zero in the pure metal Ca. The Ca is combined with more oxygen in CaO than in the pure metal.

    f. The nitrogen in N_2 is more oxidized than the nitrogen in ammonia, NH_3, because it is combined with less hydrogen. The oxidation number is zero for nitrogen in N_2 but -3 in ammonia since each H has an oxidation number of +1 .

5. There are two common oxides for hydrogen: water H_2O, and hydrogen peroxide, H_2O_2. The more common hydrogen oxide is water.

7. Oxidation refers to a reaction in which electrons are lost, hydrogen is lost, or oxygen is gained by another element. Combustion is a reaction in which oxygen combines with another element with heat given off rapidly.

9. b. CH_3CHO, The acetaldehyde, CH_3CHO, is more oxidized because it has less hydrogen than the ethanol, CH_3CH_2OH,.

11. a. oxidized The potassium atom lost an electron when it formed the K^+ ion. The oxidation number went up from zero to +1.

13. The earth's atmosphere is roughly 20% oxygen. Oxygen in the atmosphere oxidizes other substances when the oxygen gains electrons to form O^{2-} ion.

15. The iron in Fe_2O_3 is oxidized relative to the element. In the element the oxidation number is zero but in Fe_2O_3 the oxidation number is +3. The three oxygens have a combined oxidation number of $-6 = 3 \times -2$. This requires a +6 for the two iron, Fe's. Each Fe must have an oxidation number of +3.

17. Magnesium is oxidized in the Mg^{2+} ion because Mg metal has a zero oxidation number. The element loses electrons when it forms the Mg^{2+} ion.

19. The Zn^{2+} ion is oxidized because Zn metal loses electrons when it forms the Zn^{2+} ion.

21. a. $NaBH_4$ is reducing because it has the ability to donate hydrogen.
    b. Methane, CH_4, is reducing because it has the ability to donate hydrogen.
    c. Sulfur trioxide, SO_3, is oxidizing because it has the ability to donate oxygen.
    d. Potassium permanganate, $KMnO_4$, is oxidizing because it can donate oxygen.
    e. Depending on the reaction benzene, C_6H_6, can be an oxidizing agent because it can accept hydrogen and be converted to a saturated hydrocarbon, C_6H_{12}. On the other hand benzene can burn and donate hydrogen to oxygen. The combustion would oxidize the benzene and reduce oxygen, making the benzene a reducing agent.

23. The compound with the higher ratio of oxygen to carbon is more oxidized. Glucose, $C_6H_{12}O_6$, has a ratio of one oxygen atom per one carbon, 1 O atom / 1 C atom. Carbon dioxide, CO_2, has a ratio of two oxygen atom per one carbon atom, 2 O atoms / 1 C atom. The CO_2 is more oxidized.

25. Electrolysis of water reduces H^+ at the cathode to produce $H_2(g)$. At the anode, oxide, O^{2-}, in the H_2O molecule is oxidized to form oxygen gas, $O_2(g)$. See page 307.

27. Corrosion of iron requires oxygen, O_2; water, H_2O; and iron, Fe. The rusting of automobiles is less of a problem in Arizona because the air is less humid there than in Chicago. There is less water in the air in Arizona.

29. The oxide coating on aluminum is tightly bonded to the surface and protects the Al by preventing additional oxidation. The oxide coating that makes up the rust on iron is not bonded to the surface; therefore, it flakes off and does not protect the underlying iron.

31. Lithium, Li, is the strongest reducing agent. It gives up electrons more readily than any other molecule, atom or ion in the table. The weakest reducing agent is fluoride ion, F^-.

33. The lead storage battery is used in automobiles. The Ni-Cad battery is used in portable video cameras. The lithium ion battery is used in cellular phones.

35. Oxidation occurs at the anode where electrons are produced. The electrons travel through an external circuit to the cathode where they are picked up by the substance that is reduced. Electricity flows internally through the battery because ions can flow through the salt bridge that connects the anode and cathode compartments.

37. Many dry cells use the anode as the reaction container. The battery produces electricity by simultaneously consuming the anode and weakening the walls of the battery. As the anode disappears the cell walls can fail and battery contents can leak out to react with surrounding metals.

39. Ni-Cad batteries contain toxic heavy metals and the batteries have a memory which decreases the charge they can store when recharged.

41. Both use oxidation-reduction to produce electrical energy. The reactants are added to a fuel cell and products are removed, while a battery is usually self-contained so that reactants and products are kept in the battery case.

43. See answer to question 41. The battery discharges to produce electricity and the contents of the battery are confined to the case of the battery. The reactants and products are still inside the battery. No material was added or removed and no mass was gained or lost.

45. Silver in tarnish is present as silver ion while the pan contains aluminum atoms. Silver ion, Ag^+, is a better oxidizing agent and Al is a reducing agent so silver ions gain electrons from aluminum atoms, forming silver atoms and Al^{3+} ions. The water-baking soda solution allows transfer of electrons.

47. Aluminum is a more active metal than copper. The aluminum will oxidize to give electrons to reduce the copper ion, Cu^{2+}.

$$2\ Al_{(s)} +\ 3\ Cu(NO_3)_{2(aq)} \longrightarrow\ 3\ Cu_{(s)} +\ 2\ Al(NO_3)_{3(aq)}$$

Oxidation-Reduction in Chemistry

Across

2 Oxidation-reduction cell in which reactants are added from the outside.
5 An alternate way to say that a solution is basic.
9 A compound that has lost hydrogen is called _____ .
13 The tendency of a metal to lose electrons.
15 Process of depositing a metal using electrolysis.
16 Unwanted oxidation of metals.
17 Process of using electrical energy to produce a reaction.

Down

1 Strongest oxidizing agent
3 Cell that produces electricity.
4 The anode and cathode compartments are connected by a _____ bridge.
6 Oxidation is the _____ of electrons.
7 Automobiles use ___ acid batteries.
8 A disadvantage of the Ni-Cd battery when recharged.
9 The Leclanche cell also known as the _____ cell.
10 The lithium-_____ battery has high power to weight ratio.
11 Electrode where oxidation occurs.
12 Rechargeable batteries.
14 Electrode where reduction occurs.

156

Bridging the Gap
Silver Tarnish Removal by Oxidation - Reduction

Silver is a metal of great beauty and usefulness. Silver is used to make coins, jewelry, electrically conducting wire, decorative candle sticks, serving trays and fine table utensils. Many of these articles are made from sterling silver instead of pure silver. Sterling silver is an alloy made from 7.5 parts copper and 92.5 parts silver. The copper increases the toughness of the metal. Silver coins need to be more robust than cake servers and punch bowls. Coin silver like the kind used to make dimes, quarters and silver dollars, is an alloy of 10 parts copper and 90 parts silver.

Silver has a disadvantage. It tends to darken rapidly when an unattractive coat of tarnish forms on the metal surface. This brownish-black stain can be removed by rubbing the surface with silver polish but this removes some of the metal as well as the tarnish. The tarnish is mainly silver sulfide, Ag_2S. The sulfur that reacts with the silver comes from a bad smelling toxic gas, hydrogen sulfide, H_2S. This is the gas that is responsible for the bad smell that is emitted by a rotten egg. Hydrogen sulfide is emitted in some industrial processes and during the decomposition of plant and animal matter. Sulfur compounds are found in egg and mustard. Some silver tarnish results from contact with these foods. Mayonnaise is made using whole eggs so contact with it can cause silver tarnish.

An oxidation reduction reaction can be used to remove the sulfur from the tarnish and leave the silver intact. The silver in the silver sulfide, Ag_2S, can be reduced to silver metal by a more active metal. If the activity table in the chapter is checked, you can see that aluminum is a more active metal than silver. This means that aluminum metal can be oxidized to Al^{3+} and the Ag^{1+} can be reduced to Ag metal. A tarnish removal process is described below. It is similar to the one given in the text on page 291.

Equipment and materials
Aluminum foil, Pyrex™ glass container or Teflon™ lined pan large enough to hold the complete silver object, baking soda, tap water, heat resistant plastic spatula or a wooden utensil.
1. Make a baking soda solution by dissolving a heaping teaspoon of baking soda, $NaHCO_3$, in approximately one liter (roughly one quart) of water.

2. Use a microwave oven to heat the baking soda solution. The solution can be heated on a stove top burner in a Teflon™ lined pan or pot.

3. Place a piece of aluminum foil on the bottom of the bowl.

4. Be careful to avoid skin contact with the hot solution. Submerge the tarnished object completely in the solution.

5. Be sure the tarnished object is in contact with the aluminum foil. To speed up the process, use the wooden utensil to press the tarnished item against the aluminum. Notice that a reaction occurs almost immediately and gas bubbles form.

6. When the tarnish has been removed or there is no longer any significant change, remove the item from the solution and rinse it with clean tap water. Wipe the item dry with a soft cloth.

Bridging the Gap
Concept Report Sheet

Name _____

Silver Tarnish Removal by Oxidation - Reduction

The combination of silver sulfide, aluminum metal and soda solution make an electrolytic cell. The aluminum is the negative electrode and it is oxidized to aluminum ions. The silver is the positive electrode where the silver ions are reduced to silver atoms. There is a passage of electricity and simultaneously some hydrogen gas is produced by the reduction of hydrogen in water to hydrogen gas.

1. Nature of silver tarnish and Ag in sterling silver
    What is the formula for tarnish on silver?

    What is the oxidation number for silver ion in tarnish?

    What is the oxidation number for silver in sterling silver?

    Is silver reduced when tarnish is converted to silver and the sulfide ion is carried into solution by H^{1+} ions from the baking soda? Explain.

2. Nature of aluminum in foil and after formation of aluminum sulfide
    What is the oxidation number for aluminum atoms in aluminum foil?

    What is the formula for the positive ion formed when aluminum metal is oxidized?
    (Hint: Remember aluminum is in Group IIIA.)

    What is the formula for the sulfide ion?
    (Hint: Sulfur is in GroupVIA and has six valence electrons.)

    Write the formula for aluminum sulfide.

3. **Equation for the oxidation-reduction reaction for tarnish removal**
    Write the reaction for the conversion of silver tarnish and aluminum metal to silver metal and aluminum sulfide.

4. **Prediction for tarnish removal**
    Copper tarnish is also a sulfide, CuS(s). Do you think copper tarnish can be removed using the same procedure used to remove tarnish from silver? Justify your answer.

Solution to Speaking of Chemistry crossword puzzle

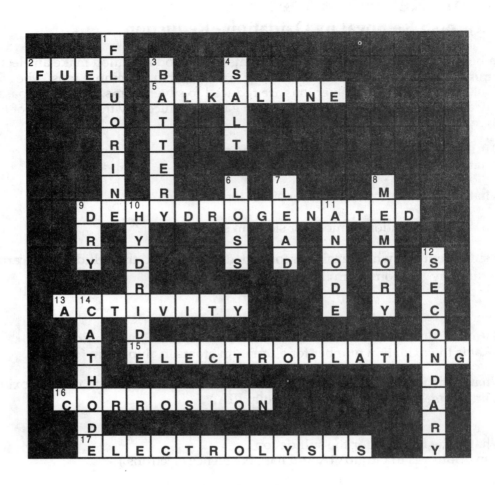

11   Chemicals from the Air, Sea, and Land

11.1   The Whole Earth

This section describes the structure and composition of the earth's crust. Definitions are given for the three major types of rocks igneous, sedimentary, and metamorphic. Minerals are defined. The forces and processes that distribute minerals in the crust are described. Regions of the world that have high concentrations of specific minerals are named. Mineral groups found in the earth's crust are tabulated. A cross sectional diagram illustrates the relationships between the earth's lithosphere, asthenosphere, mantle and core.

Objectives

After studying this section a student should be able to:

     give a definition for ecology
     describe the atmosphere and the elements it provides
     give a definition for the hydrosphere
     give definitions for igneous, sedimentary, and metamorphic rock
     state the definition for minerals
     give an approximate value in kilometers for the thickness of the earth's crust
     state the definition for an ore
     name two major mineral groups, give a specific example of a mineral in the group

Key terms

atmosphere	hydrosphere	Earth's crust
igneous rocks	sedimentary rocks	metamorphic rocks
silicates	minerals	lithosphere
major mineral groups	relative abundance	

11.2   Gases from the Atmosphere

This section describes the fractionation process for extracting gases from air. The boiling points for atmospheric gases are given. Physical properties of oxygen, nitrogen, and noble gases are described. Definitions are given for cryogens and cryogenic surgery. Nitrogen fixation is discussed. A description of the Haber process includes detailed illustrations and a discussion of the economic importance of ammonia production.

Objectives

After studying this section a student should be able to:

     describe the fractionation of air
     explain why compressed air cools when it expands
     arrange the following gases in order of decreasing boiling point: Kr, Ar, Ne, Xe, He
     give two uses for helium and explain why helium is not recovered from liquid air
     give two applications for liquid oxygen, LOX
     define cryogen and cryosurgery
     describe two hazards posed by cryogens
     explain why LOX is more hazardous than most cryogens
     tell what the Haber process does and why it is economically important
     give definitions for "noble gases" and an inert atmosphere
     give a definition for the "bends"

Key terms

fractionation of air	liquefaction	LOX
cryogens	cryosurgery	liquid nitrogen
liquid oxygen	nitrogen fixation	Haber process
noble gas	inert atmosphere	bends

11.3   Chemicals from the Hydrosphere

This section describes the substances that can be extracted from the hydrosphere. The chloralkali process to produce chlorine and sodium hydroxide from brines is described. The production of magnesium metal from sea water is diagrammed. The reactions in the electrolysis of sea water are discussed in detail.

Objectives

After studying this section a student should be able to:

define brines and salts
tell what the chloralkali industry produces and describe the chloralkali process
state what raw materials the chloralkali industry uses
explain why mercury cathodes in the chloralkali process posed an environmental hazard
tell what PVC means
define the term alloy and give two examples of low-density alloys
name two advantages offered by magnesium alloys
explain how precipitation is used to separate magnesium from sea water
tell how $CaCO_3$ from seashells is used to make lime, CaO

Key terms

brines	salt	chloralkali industry
PVC	polyvinyl chloride	electrolysis of aqueous NaCl
alloy	electrochemical cell	precipitation
insoluble	lime	calcium carbonate

11.4   Metals and Their Ores

This section describes common the connection between minerals, ores and refined metals. Iron production and the function of the blast furnace to reduce iron oxides is discussed. Compounds and reactions in the reduction process are identified. Steel production is described. The history of aluminum refining is recounted. Reactions in the Hall-Heroult process are discussed and diagrams for the electrolysis cells used in aluminum production are diagrammed.

Objectives

After studying this section a student should be able to:

state a definitions for free elements, slag, alloy, steel, ores, pig iron, cast iron, bauxite
tell what a blast furnace does to a mixture of iron oxide (Fe_2O_3), coke (C), and limestone ($CaCO_3$)
tell what pig iron is
tell how cast iron differs from pig iron
describe how steel differs from cast iron
tell why aluminum is corrosion resistant
explain why aluminum is expensive
describe the Hall-Héroult process
explain the role of molten cryolite, Na_3AlF_6, in purifying aluminum

Key terms

free elements	reactive metals	sulfides
oxides	chemical reduction	iron oxide
blast furnace	silica	slag
pig iron	cast iron	cementite
steels	carbon steel	basic oxygen process
refractory	municipal solid waste	corrosion resistant
Hall-Heroult process	bauxite	aluminum oxide
cryolite	electrolysis	

11.5   Recycling: New Metals for Old

This section describes the problems associated with disposal of materials in landfills. The factors that affect the effectiveness and success of metals recycling are discussed. The recycling of metals like lead, copper aluminum zinc, iron and steel is discussed. Green design is explained.

Objectives

After studying this section a student should be able to:

describe how the amount of municipal solid waste has grown from 1960 to 1990 (Fig.11.10)
tell which metal is the most recycled
state the major sources for recycled steel and iron
name the five most produced and valuable metals
describe the merits of waste minimization instead of collecting, processing, and recycling
explain what green design means

Key terms

solid waste	recycling plants	lead
recycled materials	market value	life span of products
cost of collection	cost of reprocessing	cost of disposal
environmental impact	quality of metal	factors in metal recycling
green design	remanufactured	

11.6   From Rocks to Glass, Ceramics, and Cement

This section describes the structure for silica, SiO_2, and silicates. The structures for the simplest network silicates (pyroxenes) and amphibole (asbestos) are illustrated. The types of asbestos and the hazards they pose are discussed in detail. Glasses and amorphous materials are defined. Some reactions in glass production are given. Ceramics are defined, the glass manufacturing process and annealing are described. The properties and applications of ceramics are described. Definitions are given for cement and concrete. The production and composition of Portland cement are described.

Objectives

After studying this section a student should be able to:

give a definition for silicates and silica
sketch a tetrahedron
name the two types of asbestos and tell which is more hazardous
state the definition for clays
give a definition for amorphous substances
state the definition for a glass
describe the purpose and reason for annealing glass

state a definition for ceramics and tell why they are economically attractive
describe composite materials
give a definition for cement
tell how mortar differs from concrete
give a definition for Portland cement

Key terms

silica	pyroxenes	tetrahedra
amphibole	fibrous natural silicates	crocidolite
asbestos	chrysotile	EPA
glass	amorphous	annealing
Pyroceram	ceramics	ceramic composites
cement	Portland cement	concrete

Additional Readings

Bailey, Jeff. "The Recycling Myth" Reader's Digest July 1995: 105.

Baker, Beth. "Curbing Recycling Revisionists: Refuting a Bevy of Critics, Recycling Advocates Say Curbside Recycling Makes Economic Sense" Environmental Action Magazine Summer 1995: 29.

Bowen, H. Kent. "Advanced Ceramics" Scientific American Oct. 1986: 168.

Jackson, James O. "World-class Litterbugs" Time Oct. 18, 1993: 80.

"Second Time Around (A Survey of Waste and the Environment)" The Economist May 29, 1993: W8.

Wallach, Jeff. "Self-healing Concrete" Popular Science Feb. 1993: 19.

Wolkomir, Richard. "Would You Want to Live in This Dump?" Smithsonian April 1994: 121.

Answers to Odd Numbered Questions for Review and Thought

1.   a. The atmosphere is the thin envelope of gases extending from the Earth's surface out into space.
     b. The hydrosphere is the layer of fresh water and salt water above and below the Earth's surface
     c. rock formed by solidification of molten rock
     d. rock formed by deposition of dissolved or suspended substances

3.   a. Cryogens are substances that have temperatures in the range of -180°C.
     b. Cryosurgery uses liquid nitrogen to remove tissue by freezing it.
     c. Nitrogen fixation is the process of converting atmospheric nitrogen into water soluble compounds that can be absorbed through plant roots.
     d. The "bends" describes a condition resulting from rapid decompression after use of pressurized gases for breathing. The rapid decompression produces "bubbles" of nitrogen from the N_2 that was dissolved in the blood at high pressure. The "bubbles" in the capillaries block circulation and can be lethal.
     e. Brines are salty water or salt solution.

5. a. Annealing is the process of heating and then slowly cooling a substance to make the substance less brittle and reduce strain in the solid.
    b. Amorphous solids have no regular crystalline order.
    c. Ceramics are materials generally made from clays and then hardened by heat.
    d. Cement is a substance able to bond mineral fragments into a solid mass.
    e. A glass is a hard, noncrystalline substance with random (liquid-like) structure.

7. The average kinetic energy for all four of the gases is the same because it depends on the temperature of the gas. The equation for kinetic energy, KE = $\frac{1}{2}[mv^2]$, shows that when the mass, m, is small the velocity, v, is high. When the mass is high the velocity is low. The gases H_2 and He have low masses and therefore, high velocities; their velocities are generally greater than the "escape velocity" needed for them to leave Earth's gravitational field. The gases N_2 and O_2 are much heavier molecules, their velocities are low, and they don't escape as readily.

9. a. liquefaction of air
    b. liquefaction of air
    c. separation from natural gas
    d. liquefaction of air

11. A refrigerated boxcar cooled by liquid nitrogen might be dangerous to enter because the boiling liquid nitrogen might displace the oxygen in the air. The atmosphere might be exclusively N_2 and contain no O_2 to breathe.

13. Magnesium, see page 320, is extracted from sea water. It is used for alloys for auto and aircraft parts, fireworks, flashbulbs.

15. Gold, copper, and platinum all can be found in the pure metallic element in nature.

17. The slag is lower density than the molten iron. The low density molten slag floats on the higher density molten iron. The two layers are not soluble in each other so they do not mix.

19. Excess carbon is used when Fe_2O_3 is reduced to iron, Fe. This excess carbon remains in pig iron as an impurity. Pig iron is brittle partly because it contains Fe_3C.

21. a. $4\ Al_{(s)}\ +\ 3\ O_{2(g)}\ \longrightarrow\ 2\ Al_2O_{3(s)}$
    b. $3\ Na_{(s)}\ +\ AlCl_{3(s)}\ \longrightarrow\ Al_{(s)}\ +\ 3\ NaCl_{(s)}$

23. bauxite

25. At the time aluminum was extremely valuable. It has low density so the cap was easily lifted into place. Aluminum reacts with oxygen in the air to form a white oxide coating that would color match the marble color of the rest of the structure and that would protect the rest of the aluminum cap from more corrosion.

27. The structure of an SiO_4 unit is tetrahedral.

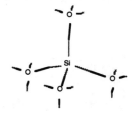

29. The two common forms of natural fibrous silicates are amphiboles (crocidolite) and serpentine (chrysotile). Both are also called asbestos. The crocidolite fibers are small needles and can fit into lung passages. They are slightly soluble and hence persistent in lung tissue. They can cause lung cancer and other health problems.

31. oxygen and aluminum

33. Annealing reduces the brittleness of a material. It produces more uniform bonding forces and minimizes strain in the structure so annealed glass is stronger.

35. It is a mixture of ceramic materials or of ceramic fibers and plastics. This makes a light-weight, heat-tolerant, high-strength material for such things as turbine blades.

37. It is a substance produced by roasting a powdered mixture of chalk, sand, clay, iron oxide, and gypsum. Typical Portland cement composition is 60-67% CaO, 17-25% SiO_2, 3-8% Al_2O_3, up to 6% Fe_2O_3 and small amounts of other ingredients.

39. Glass is noncrystalline and amorphous, while sodium chloride has a regular crystalline structure.

41. Common window glass is a mixture of $Na_{2n}(SiO_3)_n$ and $Ca_n(SiO_3)_n$.

43. Sodium, Na, is oxidized. The sodium loses electrons and changes from Na metal with zero charge to Na^{1+} ions in NaCl. Aluminum, Al, is reduced. It gains three electrons and is reduced from Al^{3+} to Al atoms.

45. Concrete is not uniform and is a mixture of cement, sand, and aggregate.
Mortar is uniform and is a mixture of cement, sand, water, and lime.
Cement is an ingredient in mortar and concrete.

47. Yes. Separation of useful pure elements would require processing more material and lead to greater environmental disturbance.

49. "Green Design" considers a product's impact on environment at all stages of its life cycle. Examples: rechargeable lithium hydride batteries, bags of laundry detergent to refill original detergent boxes, concentrated forms of laundry detergent with most inert "fillers" left out.

Speaking of Chemistry

Name _____

Chemicals from Land, Air and Sea

Across

2 Mineral used in insulation and linked to lung cancer.
5 The upper shell of the Earth including the crust and the uppermost part of the mantle.
7 A natural material that contains enough of an element to justify its mining.
8 Clays.
12 Mixture of nonmetallic waste from the refining of metals.
13 Process for the production of aluminum metal from the ore.
15 A mixture of two or more metals.
17 Third most common element in the earth's crust.

Down

1 Noble gas that is not typically recovered by fractionation of air.
2 Symbol for aluminum.
3 Low temperature liquefied gases.
4 Iron alloys.
6 Mixtures of water and salts.
8 Soluble compound formed in the Haber process.
9 Gases that typically do not form compounds.
10 Has the formula SiO_2.
11 Most abundant gas in the atmosphere with a boiling point of -196 Celsius.
14 Most recycled metal.
16 Liquid oxygen.

167

Bridging the Gap
Tetrahedron Construction and Silicates

The two dimensional illustrations in the chapter are excellent, but they cannot give a tactile sense of the tetrahedral form and the structures for extended silicates. This exercise is intended to give you a hands on experience with a three dimensional model of the tetrahedron and, if you choose, with even larger structures.

Please read all these directions before doing any cutting.

Write your name in the blank space provided on the template. Your instructor may want you to turn in your completed tetrahedron. Be careful to keep the A, B, and D tabs on the template when you cut out the tetrahedron. Be sure to leave the black edges on the faces.

Hold the cutout so you can read your name. Fold faces A, B, and D away from you. Hold face D up so you can read it. Fold the hidden support away from you. Do this same process with face B and the second hidden support. Slide the hidden support behind face A. Fold tab B over the B on face B. Insert the remaining hidden support behind face B. Fold tab A over the A on face A. Fold tab D over the D on face D. All the tabs can be secured with a piece of transparent tape if you wish. You now have your tetrahedron that represents the structure for SiO_4. The oxygens 1,2,3,4 are at the matching corners of the tetrahedron. The silicon atom is in the center of the tetrahedron. The tetrahedron is the shape for the fundamental building block of silicates.

Silicate Anions

SiO_4^{4-}  $Si_2O_7^{6-}$ $Si_3O_9^{6-}$

Silicate Chains and Asbestos

The chain structures for pyroxenes and amphiboles can be built by joining tetrahedra through their corners so an oxygen is shared between two silicon atoms in neighboring units. The chains give a fibrous material. The crocidolite double chain leads to long thin needles that are able to penetrate the airways in the lung. The crocidolite asbestos is less soluble and persists longer in tissue. This increases the risk of lung cancer and other health problems. You and your classmates can duplicate the crocidolite structure by sharing your tetrahedra.

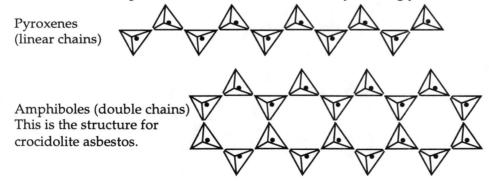

Pyroxenes (linear chains)

Amphiboles (double chains)
This is the structure for
crocidolite asbestos.

Bridging the Gap
Template for silicate tetrahedra

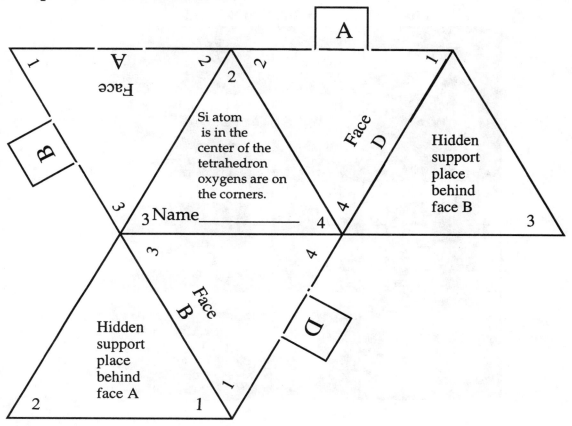

Si atom is in the center of the tetrahedron oxygens are on the corners.

Name_____

Hidden support place behind face B

Hidden support place behind face A

Solution to Speaking of Chemistry crossword puzzle

Chemicals from Air, Sea & Land

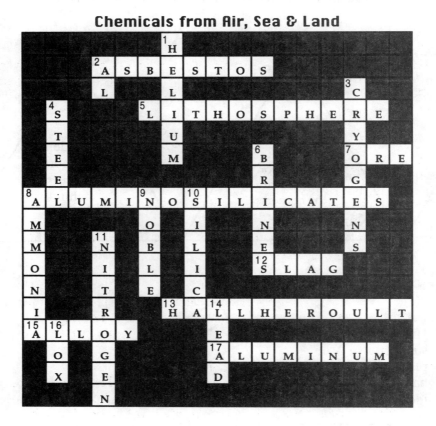

12 Energy and Hydrocarbons

12.1 Energy from Fuels

Combustion reactions are defined in this section. The origin of energy changes in combustion reactions is described in terms of the differences in bond energies for reactants and products. The calculation of heat of combustions are illustrated using bond energies. Energy values per gram are given for common fuels.

Objectives

After studying this section a student should be able to:

    give definitions for the calorie and the joule

    give a definition for bond energy

    tell what determines whether a reaction takes in or gives off heat

    give a definition for heat of combustion

    calculate the energy change for a reaction using bond energy tables

    name the four classes of hydrocarbons

Key terms

fuel	calorie	joule
hydrocarbons	combustion calorimeter	heat of combustion
average bond energies	energy per gram	alkanes
alkenes	alkynes	aromatics

12.2 Alkanes--Backbone of Organic Chemistry

Alkanes (saturated hydrocarbons) are defined in this section. The importance of the tetrahedral shape around carbons in alkanes is reviewed. Detailed explanations of structural formulas, ball and stick models, and space filling models are given. Straight and branched chain alkanes are defined. The differences between straight and branched chain alkane isomers are illustrated.

Objectives

After studying this section a student should be able to:

    give a definition for saturated hydrocarbons and alkanes

    write the general formula for alkanes

    sketch the tetrahedral structure for methane and give the bond angles

    describe how boiling points for alkanes change with increased carbon chain length

    sketch a structural formula, ball-and-stick formula and space-filling model for methane and
        for ethane

    tell how straight-chain isomers differ from branched-chain isomers

    give a definition for isomers and give examples

    write the structural formula for the isomers of C_4H_{10}

    give names for common alkanes

    tell what is meant by the term alkyl group

    give structures for the following alkyl groups:

        methyl, ethyl, propyl, isopropyl, butyl, t-butyl

    describe the naming system for alkyl groups and how they relate to alkane names

Key terms

alkane	general formula C_nH_{2n+2}	carbon-carbon single bond
saturated hydrocarbons	tetrahedral	ball-and-stick model
space-filling model	straight-chain	branched-chain
structural formula	isomers	structural isomers
methyl group	alkyl groups	alkane names
alkyl group names	"-ane" suffix	"-yl" suffix

12.3 Alkenes: Reactive Cousins of Alkanes

Alkenes are defined and their economic importance is described in this section. The general formula, C_nH_{2n}, for alkenes is given. The reactivity of alkenes is described. Structural isomers and stereoisomers are defined. The *cis* and *trans* isomers for alkenes are defined and illustrated.

Objectives

After studying this section a student should be able to:
give the definition for alkenes
write the general formula for alkenes with one double bond
draw the structural formulas for ethene and for propene
tell what bond angles are characteristic for alkenes and sketch examples of alkenes
describe *cis-* and *trans-* isomers and give examples

Key terms

alkene	general formula C_nH_{2n}	carbon-carbon double bond
alkene names	numbering the carbon chain	naming structural isomers
"-ene" suffix	propylene	structural isomers of alkenes
cis isomer	*trans* isomer	stereoisomerism
optical isomerism	free rotation	degree of flexibility of bonds

12.4 Alkynes

Alkynes are defined in this section. The general formula for alkynes, C_nH_{2n-2}, is given. The structural characteristics of alkynes are illustrated. The unusual metallic properties of polyacetylene are discussed.

Objectives

After studying this section a student should be able to:
give definitions for alkynes
write the general formula for alkynes with one triple bond
draw the structural formula for ethyne
tell what bond angle is characteristic of an alkyne

Key terms

alkynes	general formula C_nH_{2n-2}	$-C \equiv C-$
"-yne" suffix	naming alkynes	carbon-carbon triple bond

12.5 The Cyclic Hydrocarbons

Cycloalkanes are described in this section. Structures for cyclopropane, cyclobutane, cyclopentane and cyclohexane are illustrated. The chair and boat conformations for C_6H_{12} are illustrated. Aromatic compounds are defined. The carcinogenic hazards of aromatics are described. The substituted derivatives of benzene and the naming system for disubstituted isomers are explained. Polycyclic aromatics like naphthalene and anthracene are illustrated.

Objectives

After studying this section a student should be able to:

   explain how cyclic alkanes differ from straight-chain hydrocarbons

   draw structures for cyclopropane, cyclobutane, cyclopentane

   sketch the boat and chair structures for cyclohexane

   tell why the chair form for cyclohexane is more stable

   give the definition for aromatic compounds and draw the structural formula for benzene

   give the definition for carcinogens

   draw the structures for meta-, para- and ortho-dichlorobenzene, $C_6H_4Cl_2$

Key terms

cyclic hydrocarbons	aromatics	cycloalkanes
conformation	boat	chair
benzene ring	C_6H_6	planar structure
carcinogenic	delocalization of electrons	derivatives of benzene
ortho-	meta-	para-
naphthalene	anthracene	steroids
natural product chemists		

12.6   Oxygen Comes on Board

Functional groups are introduced in this section. Oxygen-containing functional groups with single bonds like alcohols and ethers are described. Wood and grain alcohol are discussed in detail. Ethers are discussed and their applications described.

Objectives

After studying this section a student should be able to:

   give the definition for a functional group

   give definitions for alcohols and ethers

   write the formulas for ethanol and methanol

   describe how methanol is produced

   describe how ethanol can be produced

   give uses for ethers

   tell what MTBE is and describe its major use

   give the definition for gasohol and two reasons for its use

Key terms and equations

functional groups	alcohols	"-ol" suffix
methanol and CH_3OH	R-OH	ethanol and C_2H_5OH
gasohol	fermentation	carbohydrates
ethers	R-O-R'	MTBE
octane enhancer	anesthetic	methyl propyl ether
diethyl ether	$C_2H_5OC_2H_5$	$CH_3OCH_2CH_2CH_3$

12.7   Petroleum

Petroleum is defined in this section. The growth in petroleum consumption is described. Fractional distillation and petroleum refining are explained and illustrated. Petroleum fractions and their uses are tabulated. Octane ratings for gasoline are explained. The reference standard, 2,2,4-trimethylpentane, for an octane rating of 100 is discussed. Octane enhancers and gasoline additives are tabulated. Catalytic re-forming of straight chain hydrocarbons is described. Reformulated and oxygenated gasolines are discussed. The catalytic cracking process is described.

Objectives

After studying this section a student should be able to:

    tell how long known oil reserves are projected to last at current consumption levels
    describe the process for refining petroleum by fractional distillation (use a sketch)
    give a definition for a petroleum fraction
    tell what is meant by straight-run gasoline
    give the definition of octane rating
    name the compound used to define the 100 octane rating
    tell how the octane rating for a fuel is determined
    explain why knocking and pinging occur in an automobile engine
    explain why catalytic re-forming of hydrocarbons is done
    describe the catalytic cracking process
    give definitions for: octane enhancers, reformulated gasoline, oxygenated gasoline
    tell why alkenes and aromatic compounds are added to straight-run gasoline

Key terms and equations

petroleum	crude petroleum	fractional distillation
petroleum fractions	distillation tower	volatile
lower-boiling fractions	higher-boiling fractions	condense
residual oil	octane rating	"straight-run"
"knocking"	"pinging"	isooctane
premium gasoline	regular plus gasoline	increasing the octane rating
octane enhancers	tetraethyl lead	catalytic re-forming process
ethanol	toluene	straight-chain hydrocarbons
MTBE	methanol	branched-chain hydrocarbons
oxygenated gasoline	reformulated gasoline	catalytic cracking process
cracking	sulfur in gasoline	benefits of oxygenated gasoline

12.8   Natural Gas

The composition of natural gas is described in this section. Current use and projected U.S. consumption of natural gas are discussed. Efforts to use natural gas as a vehicle fuel are described.

Objectives

After studying this section a student should be able to:

    describe the composition of natural gas
    describe the uses for natural gas
    explain the benefits and disadvantages associated with using natural gas as a vehicle fuel

Key terms and equations

natural gas	C_1 to C_4 alkanes	methane
liquefied gas	butane	propane
benefits of vehicle fuel	organic chemical industry	energy source

12.9   Coal

The composition of coal is described in this section. The current levels of coal production and coal reserves are discussed. Coal gasification is explained using illustrations and reactions for two different processes. Coal liquefaction is described.

Objectives
After studying this section a student should be able to:
        describe the composition of coal
        tell why the use of coal decreased as a heating fuel
        tell how synthesis gas is produced
        describe the coal gasification process
        explain how coal gasification differs from coal liquefaction

Key terms and equations

coal	mixture of hydrocarbons	fused rings of carbon atoms
deep coal mining	strip mining	coal reserves
coal gasification	synthesis gas	environmentally clean fuel
energy-efficient	coal liquefaction	hydrogenating the coal

12.10   Methanol as a Fuel
This section describes programs to replace gasoline with methanol and other fuels. Flexible-fueled vehicles are defined. Definitions are given for alternate fuels and mixtures. M100 is pure methanol and M85 is a blend of 85% methanol with 15% gasoline. Advantages and disadvantages of methanol and alternate fueled vehicles are discussed.

Objectives
After studying this section a student should be able to:
        give reasons for using methanol as a replacement fuel
        explain why FFVs are being developed
        tell what the abbreviation code names for fuels like M100, E100, and M85 mean
        give two advantages that methanol offers as a fuel
        describe two problems that methanol presents as a fuel

Key terms and equations

methanol	M100	M85
ethanol	E100	E85
FFV, flexible-fueled vehicles	exhaust emissions	

Additional Readings
Anderson, E.V. "Ethanol's Role in Reformulated Gasoline Stirs Controversy" <u>Chemical & Engineering News</u> Nov. 2, 1992: 7.

Anderson, E.V. "Health Studies Indicate MTBE Is Safe Gasoline Additive" <u>Chemical & Engineering News</u> Sept. 20, 1993: 9.

Bowler, Sue. "Where the Power Lies" <u>New Scientist</u> Jan. 23, 1993: 32.

Emsley, John. "Energy and Fuels" <u>New Scientist</u> Jan. 15, 1994: 51.

Kerr, Richard A. "U. S. Oil and Gas Fields Double in Size" <u>Science</u> Feb. 24, 1995: 1090.

Lewis, Bernard . "High Temperatures: Flame" <u>Scientific American</u> Sept. 1954: 84.

Maitlis, Peter & John Rourke. "Rich Seams for Chemicals" <u>New Scientist</u> Jan. 23, 1993: 37.

Peaff, George "Court Ruling Spurs Continued Debate Over Gasoline Oxygenates" <u>Chemical & Engineering News</u> Sept. 26, 1994: 8.

Rogge, Wolfgang F. et al. "Natural Gas Home Appliances (ES &T Research: Sources of Fine Organic Aerosol, part 5)" <u>Environmental Science & Technology</u> Dec. 1993: 2736.

Spencer, Brian. "Hydrocarbon Fuels, No Thanks" <u>New Scientist</u> Mar. 18, 1995: 45.

Answers for Odd Numbered Questions for Review and Thought

1. Fossil fuels are coal, crude oil, natural gas or heavy oil formed over millions of years by the action of heat and pressure in the earth's crust on buried decayed animal and plant matter.

3. An hydrocarbon is a compound such as an alkane, alkene or alkyne containing only carbon and hydrogen.

5. The heat of combustion is the energy released when a compound reacts completely with oxygen.

7. a. A multiple bond exists between two atoms that share either four or six electrons.

   b. A double bond exists between two atoms sharing four electrons. An example is the bond that exists between the carbon atoms in ethylene, CH_2CH_2.

   c. A single bond exists between two atoms sharing two electrons. An example is the bond between carbon atoms in ethane, CH_3CH_3.

   d. A triple bond exists between two atoms sharing six electrons. An example is the bond between carbon atoms in acetylene, H-C:::C-H

9. An alkyl group is a hydrocarbon fragment formed by removing a hydrogen atom from an alkane.

11. All angles in methane, CH_4, are 109.5° regardless of the way the molecule is illustrated.

   or

13. Alkanes have the formula C_nH_{2n+2}. Alkyl groups have the formula C_nH_{2n+1} with one less hydrogen atom.

15. a. isopropyl, CH_3CHCH_3

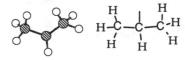

   b. butyl, $CH_3CH_2CH_2CH_2$

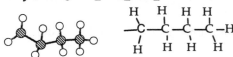

178

c.   t-butyl, $(CH_3)_3C$

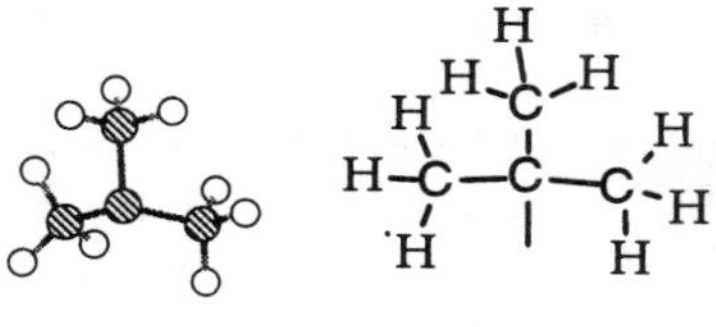

17.   The numbering sequence starts at the carbon atom that will give the lowest integer for the substituents.

a. 2,2,4-trimethylhexane

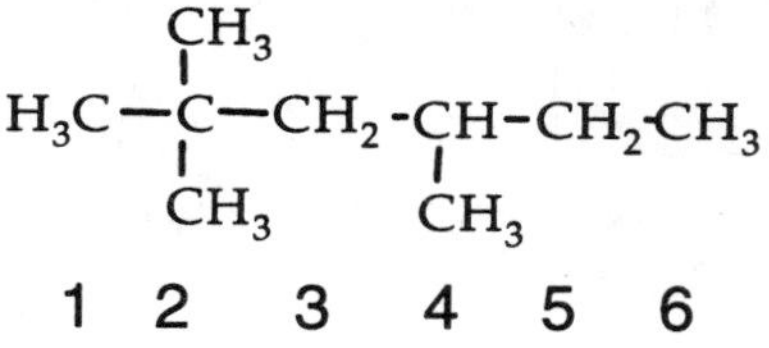

$$H_3C-\underset{\underset{CH_3}{|}}{\overset{\overset{CH_3}{|}}{C}}-CH_2-\underset{\underset{CH_3}{|}}{CH}-CH_2CH_3$$

1   2   3   4   5   6

b.   3-ethylpentane

$$H_3C-CH_2-\underset{\underset{CH_3}{\overset{|}{\overset{H_2C}{|}}}}{CH}-CH_2-CH_3$$

1   2   3   4   5

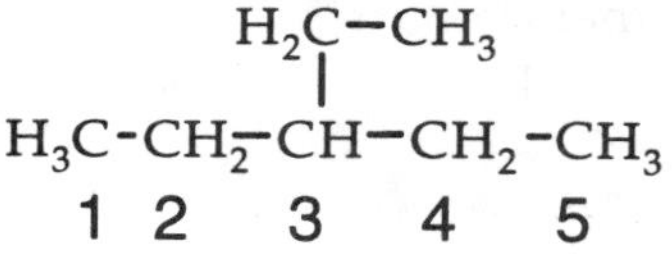

c.   Only one structure possible so it is called methylbutane not 2-methylbutane.

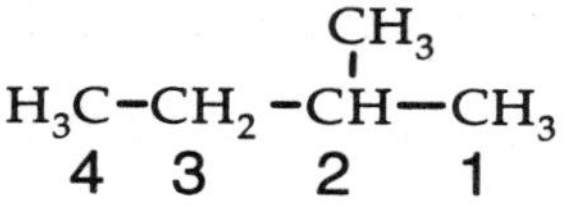

$$H_3C-CH_2-\underset{\overset{|}{CH_3}}{CH}-CH_3$$

4   3   2   1

19.   There are three structural isomers for C_4H_{10}. There is no simple way to determine the number of such isomers. The only way to do this is to follow a systematic set of steps. The first thing to do is to draw the structure for the straight chain isomer. The next is to shorten the chain by one carbon and attach this carbon to the number two carbon on the chain. The "moving" or "attaching" carbon atom is then positioned on the next carbon and so forth until duplicate structures are produced. After all these possibilities are exhausted the carbon chain is shortened by another carbon atom and the possible combinations are again tried. There are only the following three different butane isomers.

$CH_3CH_2CH_2CH_2CH_3$ $CH_3\underset{\overset{|}{CH_3}}{CH}CH_2CH_3$ $CH_3\underset{\underset{CH_3}{|}}{\overset{\overset{CH_3}{|}}{C}}CH_3$

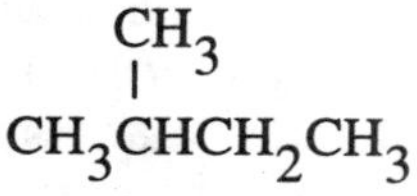

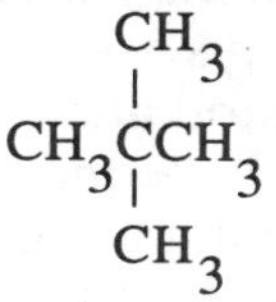

21.   There are five structural isomers for C_6H_{14}.

$CH_3CH_2CH_2CH_2CH_2CH_3$ hexane

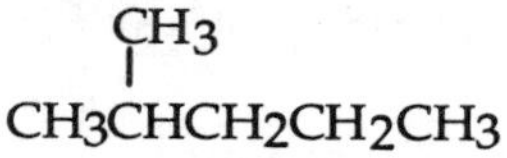

$CH_3\underset{\overset{|}{CH3}}{C}HCH_2CH_2CH_3$ 2-methylpentane

$CH_3CH_2\underset{\overset{|}{CH3}}{C}HCH_2CH_3$ 3-methylpentane

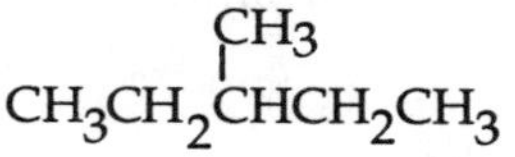

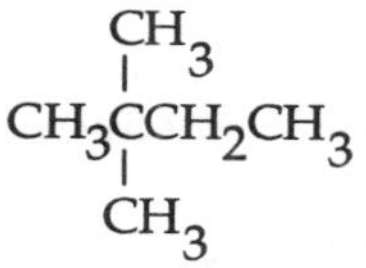

$CH_3\underset{\underset{CH_3}{|}}{\overset{\overset{CH_3}{|}}{C}}CH_2CH_3$ 2,2-dimethylbutane

$CH_3\underset{\underset{CH_3}{|}}{\overset{\overset{H_3C}{|}}{C}}H\underset{\overset{|}{CH_3}}{C}HCH_3$ 2,3-dimethylbutane

23. The structure for 2,3,3-trimethylpentene can be drawn either way shown. Note the numbering sequence gives the lowest number for the double bond.

$$
\underset{1\quad\; 2\quad\; 3\quad 4\quad 5}{H_2C{=}\underset{\underset{CH_3}{|}}{C}{-}\overset{\overset{CH_3 CH_3}{|\;\;|}}{C}{-}CH_2{-}CH_3}
\qquad or \qquad
\underset{1\quad\; 2\quad\; 3\quad 4\quad 5}{H_2C{=}C{-}\underset{\underset{CH_3\;\,CH_3}{|\quad\;|}}{\overset{\overset{CH_3}{|}}{C}}{-}CH_2{-}CH_3}
$$

25. The two substituents are on the same side of the double bond in *cis-*dichloroethene.

The two substituents are on opposite sides of the double bond in *trans-*dichloroethene

27. Cyclobutane, C_4H_8, can be dawn as a square, figure A. This correctly shows the equivalency of the four carbon atoms. The flat drawing does not give an accurate picture of the tetrahedral shapes around each carbon. The actual molecule is shaped more like a square that has been bent, figure B. The atoms take positions in three dimensions that minimize the energy of the structure.

Figure A   Figure B

29. Crude oil is heated to about 400°C. The mixture of hot vapors is passed into a fractionation tower. As the hot vapors rise in the tower, they cool. When they cool to their condensation temperatures (boiling temperature), they condense. This means each compound in the vapor mixture will condense at some specific height in the tower. Collection trays are stacked in the tower. The temperatures of the collection trays are cooler at the top and hotter at the bottom of the tower. The compounds with small molecular masses tend to have low boiling points and are collected nearer the top of the tower. The compounds with high molecular masses tend to have high boiling points and are collected near the bottom. This is why compounds with extremely high molecular weights are called "bottoms". Products are really collected based on their boiling points.

31. Straight-run gasoline is a hydrocarbon mixture with a boiling point range of 30-200 °C. The molecules are hydrocarbons that have five to twelve carbon atoms, C_5-C_{12}.

33. Gasohol is defined by law. It is a blend of 90% hydrocarbon and 10% ethanol.

180

35. a. Catalytic re-forming is used to raise the octane rating of straight-run gasoline. The straight chain hydrocarbons are rearranged into branched chains. Some compounds are converted to aromatic hydrocarbons. Both branched hydrocarbons and aromatics have higher octane ratings.
b. Catalytic cracking is used to break long chain hydrocarbons into shorter-chain hydrocarbons. These products are a mixture of alkanes and alkenes. These compounds have higher octane ratings. The alkenes can also be used as starting materials (feed stock) for other chemical products.
c. The octane rating of a specific fuel equals the percentage of isooctane in an isooctane-heptane mixture that gives the same knocking and pinging behavior as the specific fuel.

37. Higher boiling hydrocarbons are usually branched hydrocarbons.

39. Octane ratings are determined by using the specific fuel to be tested in a standard engine. The standard engine is run using the test fuel and its knocking properties are compared with those of isooctane and heptane mixtures. The percentage of isooctane in the matching mixture equals the octane for the test fuel.

41. Oxygenated gasolines burn more completely. This is expected to reduce hydrocarbon exhaust emissions, carbon monoxide levels and ozone emissions.

43. Treating pulverized coal with superheated steam forms either carbon monoxide and hydrogen or, using a catalyst, carbon dioxide and methane.

45. The M identifies a fuel blend of methanol and gasoline. The E identifies a fuel blend of ethanol and gasoline. The number following the symbol tells the percentage of these alcohols in the mixture. The remaining component in the blend is gasoline.
M100 means 100% methanol.
E100 means 100% ethanol.
M85 means 85% methanol, 15% gasoline.
E85 means 85% ethanol, 15% gasoline.
FFV means flexible-fueled vehicle which can operate on gasoline or other fuels.

47. a. alkane   b. alkyne   c. ether   d. alcohol   e. aromatic

49. Lead is a toxic heavy metal. Lead emissions contaminate the atmosphere, soil, and ground water. Lead poisoning can be fatal to fish, animals and humans. There is evidence that brain damage and other problems can result from prolonged exposure to lead levels that are below the toxic level.

51. Addition of aromatic compounds to gasoline would raise octane ratings, but there are many disadvantages. Aromatics are typically carcinogens. Evaporated fuel, spilled fuel and exhaust emissions would contain more carcinogens, causing greater human cancer risks. These compounds are also implicated in the production of photochemical smog. This would increase the social cost in terms of respiratory diseases.

Solutions for Selected Problems

1. a. The balanced equation is $2\ C_4H_{10}\ +\ 13\ O_2\ \longrightarrow\ 8\ CO_2\ +\ 10\ H_2O$

    b. The energy change is 630 kcal/mol. The following method works to determine this energy change for the reaction. The bond energies for all the bonds broken in reactants and all the bonds formed in the products are found in the bond energy table. The ball and stick models for the reactants and products help to identify the types of bonds involved.

$$2\ C_4H_{10}\ +\quad 13\ O_2\quad \longrightarrow\ 8\ CO_2\quad +\quad 10\ H_2O$$

The energy required to break bonds

Carbon carbon single bonds	6 x 83 kcal / mol	=	498 kcal
Carbon hydrogen bonds	20 x 99 kcal / mol	=	1980 kcal
Oxygen oxygen double bonds	13 x 118 kcal/ mol	=	1534 kcal
	Total energy taken in:		4012 kcal

The energy released is determined by the bonds formed.

Carbon oxygen double bonds	16 x 192 kcal/ mol	=	3072 kcal
Hydrogen oxygen single bonds	20 x 111 kcal/mol	=	2220 kcal
	Total energy released:		5272 kcal

The energy change for the reaction is the difference between these two total energies. 5272 kcal – 4012 kcal = 1260 kcal Since the total energy released (5272 kcal) is greater than the total energy used (4012 kcal) there is a net release of 1260 kcal when 2 mols butane react. The energy change for one mol of butane is

   1260 kcal ÷ 2 mol = 630 kcal / mol

3. a. The balanced equation is $2\ C_8H_{18}\ +\ 25\ O_2\ \longrightarrow\ 16\ CO_2\ +\ 18\ H_2O$

    b. The energy change is 1232 kcal/mol. This is calculated the same way as in problem 1.
    Total energy in: The energy required to break bonds

Carbon carbon single bonds	2 x 7 x 83 kcal / mol	=	1162 kcal
Carbon hydrogen bonds	2 x 18 x 99 kcal / mol	=	3564 kcal
Oxygen oxygen double bonds	25 x 118 kcal/ mol	=	2950 kcal
			7676 kcal

Total energy released: The energy released is determined by the bonds formed.

Carbon oxygen double bonds	2 x 16 x 192 kcal/mol	=	6144 kcal
Hydrogen oxygen single bonds	2 x 18 x 111 kcal/mol	=	3996 kcal
			10140 kcal

The energy change for the reaction is the difference between these two energies. 10140 kcal – 7676 kcal = 2464 kcal released
This is for two moles of butane. The energy change for one mol of octane is
   2464 kcal ÷ 2 mol = 1232 kcal / mol

   c. The molar mass for C_8H_{18} is needed to calculate the energy change for a gram of octane
   1 mole C_8H_{18} = 8 x 12 grams carbon + 1 x 18 gram hydrogen = 114 grams C_8H_{18}
   1232 kcal / mol octane ÷ 114 gram / mol octane = 10.8 kcal / gram octane
   1232 kcal / mol octane ÷ 114 gram / mol octane = 10.8 kcal / gram octane

182

Speaking of Chemistry

Name _____

Energy and Hydrocarbons

Across

1 Name for alkyl group, $CH_3CH_2CH_2$,

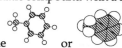 or

3 Aromatic compound with a methyl group on

benzene or

6 Abbreviation for octane enhancer, methyl-tertiary-butyl ether.

8 Hydrocarbon with formula $CH_3CH_2CH_3$.

9 Cyclohexane conformation with more crowding than

the chair form

11 Solid mixture of hydrocarbons and small amounts of sulfur.

14 Type of model of molecules that uses sticks for bonds and balls for atoms.

15 Term used to describe groups of atoms like OH in alcohols.

16 Suffix used to indicate the presence of a double bond in a hydrocarbon.

17 Dark oily mixture of hydrocarbons occurring in deposits worldwide.

19 Compounds with general formula ROR'.

20 A measure of volume equal to 42 gallons

Down

2 The prefix used to indicate the position of substituents in an aromatic molecule like.

or .

4 The functional group characteristic for alcohols.

5 Oxygen containing compound with the formula $CH_3CH_2OCH_2CH_3$.

7 Representation of molecules showing the relative size of atoms and their positions.

8 Disubstituted benzenes with groups in the 1 and 4

positions, or

10 Hydrocarbon fragment made by removing one hydrogen from an alkane.

12 Hydrocarbons with only single carbon-carbon bonds.

13 Six electrons are shared as in HC:::CH or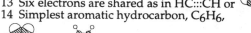

14 Simplest aromatic hydrocarbon, C_6H_6,

 or .

18 The prefix used to indicate the position of substituents in an aromatic molecule like.

 or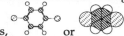

183

Bridging the Gap

Vehicle Emissions Tests: Reading the Results

A sample emissions test report is illustrated below. This type of report is required in many areas of the country before a vehicle can be licensed. This exercise is intended to familiarize you with the kinds of information that appears on one of these test forms. The emissions test looks at hydrocarbons, HC; carbon monoxide, CO; and the sum of carbon monoxide and carbon dioxide, CO + CO_2. The form shows the standards for the specific vehicle. The actual results are printed below the standard allowed emissions. The allowed HC emission value is 220 parts per million. The actual value is only 3 PPM so the vehicle passed and it is burning gas relatively efficiently. The carbon monoxide standard is 1.2%. The actual value is 00.00%. This means the vehicle is not emitting any toxic CO. The minimum level of carbon monoxide and carbon dioxide is 6.00%. The actual test result is 15.10% which is consistent with efficient burning of gasoline and a properly working catalytic converter. This vehicle passed in all three categories. This is indicated in by the big PASS opposite the final results arrow as well as the pass symbols under the separate results. A similar test report sheet appears on the following page. Your assignment is to determine whether or not the second vehicle passed. You are supposed to tell how well the vehicle met the standards, or, if it failed, tell what standards were not met. You also are supposed to tell what changes in emissions might be needed.

VEHICLE EMISSIONS TEST REPORT

This Form is Your Receipt for the Test Fee

6726078
6726078

This test was performed in accordance with federal regulations on emision tests (40 CFR 85 Subpart W)

Inspection information

License	Vehicle number	Year	Make		
ACS 139	4S2CY45V783200172	93	HONDA	FINAL RESULTS	PASS

VEHICLE INFORMATION

G.V.W.R.	ODOMETER x 1000	FUEL TYPE	TEST NO
	34	G	1

INSPECTION STATION INFORMATION

INSPECTOR.	STATION	MO. DAY YR	HR. MIN.
138	05	2 12 94	13 24

GASOLINE EMISSION TEST RESULTS

TEST: TWO SPEED	HC (PPM)	CO (%)	CO + CO_2 (%)	OPACITY(%)
Emissions Limits	0220	01.20	≥6.00	
Emissions at 0749 RPM	0003	00.00	15.10	
Emissions at 2256 RPM	0007	00.00	15.24	
	P	P	P	
RESULTS: PASS: (P) FAIL: (F)				

Bridging the Gap I
Concept Report Sheet

Name _____

Use the sample test report below to answer the following questions.

Did this vehicle pass the emissions test?

How well did the hydrocarbon emissions meet the emissions limits?

Is the engine doing a good job of completely burning the gasoline? Explain.

How well did the vehicle meet the CO emission limits? Does the catalytic converter appear to be working well?

VEHICLE EMISSIONS TEST REPORT
This Form is Your Receipt for the Test Fee

6726092
6726092

This test was performed in accordance with federal regulations on emision tests (40 CFR 85 Subpart W)

Inspection information					
License	Vehicle number	Year	Make		
ACS 719	7S2CY45V783200172	89	FORD	FINAL RESULTS	FAIL

VEHICLE INFORMATION			
G.V.W.R.	ODOMETER x 1000	FUEL TYPE	TEST NO
	64	G	1

INSPECTION STATION INFORMATION			
INSPECTOR.	STATION	MO. DAY YR	HR. MIN.
121	06	4 19 96	11 44

GASOLINE EMISSION TEST RESULTS				
TEST: TWO SPEED	HC (PPM)	CO (%)	$CO + CO_2$ (%)	OPACITY(%)
Emissions Limits	0220	01.20	≥ 6.00	
Emissions at 0749 RPM	0340	04.00	3.20	
Emissions at 2256 RPM	0422	08.00	3.80	
	F	F	F	
RESULTS: PASS: (P) FAIL: (F)				

Solution to Speaking of Chemistry crossword puzzle

Energy and Hydrocarbons

¹P	R	²O	P	Y	L		³T	⁴O	L	U	E	N	⁵E		
		R						H					T		
	⁶M	T	B	E					⁷S				H		
		H					⁸P	R	O	P	A	N	E		
	⁹B	O	¹⁰A	T			A		A				R		
		L	K			¹¹C	O	A	L		¹²A		¹³T		
		Y					R		C		L		R		
	¹⁴B	A	L	L	A	N	D	S	T	I	C	K	I		
	E						L		A		P				
¹⁵F	U	N	C	T	I	O	N	A	L		L		L		
	Z						L		¹⁶E	N	S		E		
¹⁷P	E	T	R	O	L	E	U	¹⁸M		N	S		B		
	N						E		G				O		
¹⁹E	T	H	E	R	S		T						N		
				²⁰B	A	R	R	E	L				D		

13 Alternate Energy Sources

13.1   Speaking of Energy

The concept that all energy sources produce a common product, but have different efficiencies is introduced. Energy units are defined and equivalencies between them are given. The quad is defined. Energy and power are defined and differences between the two concepts are explained. The watt, kilowatt and horsepower are described. Examples of energy unit conversions are illustrated.

Objectives

After studying this section a student should be able to:

give definitions for common energy units
convert between energy units such as joules to calories
state the definitions for power and watts
identify the five major energy sources and arrange them in order of their importance
describe the relative importance of the following energy sources in the United States
      crude oil, coal, natural gas, nuclear power, hydroelectric power

Key terms

calorie, cal	joule, J	British thermal unit, BTU
quad	power	energy equivalents
watt, W	horsepower	1 J/sec
kilocalorie, kcal		food calorie, Cal

13.2   Electricity--Energy Converted from One Form to Another

The generation of electricity by the rotation of a metal wire loop in magnetic field is described. Costs of electricity production are explained. The thermodynamic efficiency or inefficiency of electricity generation from fossil fuels and steam is described. The 1994 U.S. consumption of 27.3 quads of energy to generate 8.0 quads electricity is discussed. A 1000 megawatt coal burning power plant is used as an example to illustrate the inefficiencies of electricity generation. Examples are given showing how to calculate the number of barrels of oil and cubic feet of natural gas required to produce any number of kilowatt hours of electricity. The way to determine the kilowatt consumption of household appliances is shown. A way to calculate the cost of operating electrical appliances is illustrated.

Objectives

After studying this section a student should be able to:

give a definition for an electrical generator
give a description of how an electrical generator works
state the approximate amount of energy used each year to produce electricity in the U.S.
explain why the U.S. needed to use 27.3 quads of energy to produce only 8 quads
    of electricity in 1994
explain how to use a kilowatt hour value and determine its equivalent in barrels of oil or
    cubic feet of natural gas

Key terms

electricity
electric current
incandescent bulb
kilowatt-hours, kWh

generator
resistance
fluorescent bulb
kilowatts

second law of thermodynamics
efficiency
megawatt, MW

13.3   Nuclear Energy

This section defines nuclear fission and fusion. It illustrates and describes thermal neutrons, fission products, critical mass, and chain reactions. An historical summary of the practical use of fission is included. The source of nuclear energy from the conversion of mass to energy is explained. The relative stability of isotopes is diagrammed and explained in terms of binding energy per nuclear particle (nucleon).

Objectives

After studying this section a student should be able to:

give definitions for nuclear fission and fusion
describe the fission process and the role of thermal neutrons in fission
give definitions for fission products, critical mass, and chain reaction
describe the work done by Lise Meitener
tell why natural uranium must be enriched to support fission
state the Einstein mass energy equation
give the definitions for nucleon and binding energy per nucleon
tell how to calculate the energy yield from a change in nuclear mass
describe the origin of nuclear binding energy

Key terms

nuclear energy
fusion process
nuclear reactor
moderator
nuclear fuel
$^{238}_{92}U$

Einstein's equation

mass difference

fissionable
thermal neutrons
chain reaction
fission bomb
enriched uranium
$^{235}_{92}U$

$E = mc^2$

binding energy

fission process
fission products
critical mass
subcritical masses
nucleon
$^{239}_{94}Pu$

UF_6

binding energy per nucleon

13.4   Using Nuclear Fission and Nuclear Fusion

This section describes Enrico Fermi's work on controlled nuclear fission. It gives definitions for nuclear reactor components. The layout of a nuclear power plant is diagrammed. Significant nuclear power plant accidents such as Three Mile Island in the U.S. and Chernobyl in the former Soviet Union are described. The hazards of a nuclear reactor core meltdown are explained. Nuclear wastes are defined. Problems associated with nuclear waste storage and disposal are discussed. Nuclear fusion research, controlled fusion and plasmas are described.

Objectives

After studying this section a student should be able to:

describe the function for each of the following in a nuclear reactor:
    moderator, neutron absorber, shielding, heat-transfer fluid
describe the scope of the 1979 accident at the Three Mile Island nuclear plant
describe the scope of the 1986 accident that occurred in the Soviet reactor at Chernobyl

give a description of what is predicted to happen during a nuclear reactor core meltdown
tell why nuclear power plant wastes are a public concern
describe proposals for the safe disposal of nuclear wastes
describe the nuclear fusion process
define the terms thermonuclear and plasma
explain why a "magnetic bottle" is needed in plasma and fusion research
tell why nuclear fusion is an attractive potential energy source

Key terms

atomic reactor	shielding	neutron absorber
heat-transfer fluid	core	core meltdown
reactor fuel	fuel rod	ordinary (light) water
heavy water	Three Mile Island	Chernobyl Unit 4 Reactor
radioisotopes	beta-particle emitters	nuclear waste products
$^{137}_{55}Cs$	$^{131}_{53}I$	spent fuel rods
reprocessed	high-level wastes	unavoidable Pu production
actinides	glass logs	integral fast reactor
weapons-grade plutonium	nuclear warhead material	deuterium & tritium
fission products	underground repository	Yucca Mountain, Nevada
fusible atoms	magnetic bottle	plasma
controlled fusion	thermonuclear bomb	

13.5   Solar Energy--Almost Free

This section describes the total energy produced by the sun in a second. An estimated value for the amount of solar energy that reaches the earth is given. The fate of solar energy that the earth receives is detailed. Alternate energy sources resulting from thermal gradients that produce wind and water currents are discussed. Active and passive solar heating are described. Direct conversion of solar energy to electricity by photovoltaics is described.

Objectives

After studying this section a student should be able to:

tell the approximate % of incoming solar radiation that is reflected back into space
identify the two ways that solar energy reaching the earth's surface is utilized
tell how wind energy and ocean currents are related to solar energy and thermal gradients
describe the difference between passive and active solar heating
describe applications of photovoltaics (solar cells)
sketch and label a solar cell showing how it supplies electricity to an external circuit

Key terms

solar energy	thermal conversion	photoconversion
solar heating	winds	thermal gradients
waves	currents	Gulf Stream
solar concentrators	passive solar heating	active solar heating
endothermic	solar cell	photovoltaic cell
n-type semiconductor	p-type semiconductor	external circuit

13.6   Biofuels

This section defines biomass fuels and biofuels. Research to improve traditional biomass fuels is described. The present day production costs of fossil fuels and biomass diesel fuel are compared.

Objectives
After studying this section a student should be able to:
> give definitions for biomass fuels
> tell how photosynthesis and biomass fuels are related to carbon fixation
> describe how microalgae are used to produce biofuels
> explain why biomass diesel fuel has not replaced petroleum diesel fuel

Key terms
biomass fuels	photosynthesis	carbon fixation
microalgae	biodiesel fuel	genetic engineering

13.7   Geothermal Energy
This section describes the origin of the heat stored in the Earth's molten core. Geological hot spots are defined. Examples of practical geothermal energy projects are given. Problems with utilizing geothermal energy are described.

Objectives
After studying this section a student should be able to:
> state the range of temperatures in the Earth's core and mantle
> give reasons for the high temperatures in the Earth's core
> describe two examples where geothermal energy is a practical energy source
> name two potential problems associated with utilizing geothermal energy

Key terms
geothermal energy	tectonic plates	hot spots
core	mantle	crust
fumaroles	geothermal well	geothermal steam
hydrogen sulfide gas	superheated water	dissolved minerals

Additional Readings
Anderson, Christopher & Michael Cross. "Fusion Research at the Crossroads" Science April 29, 1994: 648.

Anderson, Earl. "Brazil's Program to Use Ethanol as Transportation Fuel Loses Steam" Chemical & Engineering News Oct. 18, 1993: 13.

Griffin, Rodman D. "Driving Out Oil Dependency...The Push To Develop Clean Cars" CQ Researcher July 10, 1992: 584.

Hoffman, Carl. "Energy Futures; Biomass: Its Cleaner Than Coal, Safer Than Nukes, More Reliable Than Solar...." Audubon Sept.-Oct. 1993: 112.

Linden, Eugene. "A Sunny Forecast: Always Cleaner Than Fossil Fuels, Renewable Power Sources May Soon Be Just As Cheap" Time Nov. 7, 1994: 66.

McGowan, Jon G. "Tilting Toward Windmills" Technology Review July 1993: 38.

Nadis, Steve. "Highway Winds: A Surprising New Alternative in the Search for Energy" Omni July 1994: 18.

Reed, Marshall J. "Geothermal Energy" Geotimes Feb. 1993: 12.

Stone, Laurie & Johnny Weiss. "The Passive Solar Home: Past Mistakes in Solar Home Design Can Be a Boon to Builders" Mother Earth News Feb.-Mar. 1995: 18.

Summers, Claude M. "The Conversion Of Energy" Scientific American Sept. 1971: 148.

Tamkins, Teresa. "Tilting at Wind Power: Wind Power Is Clean and Efficient - and It Can Kill Eagles" Audubon Sept.-Oct. 1993: 24.

"These Farms May Harvest Wind - on the Ocean" Business Week May 30, 1994: 129.

Answers to Odd Numbered Questions for Review and Thought

1. a. Renewable energy is normally replenished by natural processes, i.e. solar energy.
    b. Power is the rate of energy use; it is measured in watts (joules per second) or horse power.
    c. A Quad is equal to a quadrillion British thermal units or 1×10^{15} Btu.
    d. A thermal neutron is a relatively slow neutron with a kinetic energy about the same as that of a gas molecule at room temperature.

3. The second law of thermodynamics imposes an inefficiency in energy transfer that can be predicted mathematically. Efficiency $= \dfrac{T_{High} - T_{Low}}{T_{High}}$. Here the kelvin temperatures refer to the high operating temperature of the power plant and the low temperature for the cooling system. A typical fossil fuel power plant operates at 500°C and a cooling temperature of 27°C. This type of power plant would have a maximum efficiency of about 61 %

    $$\text{Efficiency} = \frac{773 \text{ kelvin} - 300 \text{ kelvin}}{773 \text{ kelvin}} = 0.61$$ The other steps in the power

    generating process introduce additional losses. There are mechanical and frictional losses in the turbines.
    Additional energy losses result from the mechanical losses in the generator design and losses due to the resistance of the power lines. Overall there is an efficiency of about 37.8%. This means that 62.2% of the energy produced by the original fuel is lost.

5. b. The fission of uranium-235 produces the most energy.

7. Electricity is popular because it is available on demand. You can simply click a switch to turn on electricity. It seems clean at the point of use even though the power plant may emit smoke and gases at a far distant location.

9. The fission of one mole of uranium-235 will produce more energy than the burning of a mole of gasoline.

11. The enrichment of natural uranium with additional U-235 is necessary because the natural uranium does not include a sufficient amount of fissionable uranium to produce a critical mass. There are many steps in the enrichment process. Uranium is converted to UF_6 gas. The two isotopes of uranium, U-235 and U-238, produce forms of UF_6 that have different molar masses; the molecules containing U-235 are lighter and more volatile. This volatility difference is enough to allow separation of the isotopes because the molecules containing the lighter isotope will in essence diffuse faster.

13. The energy produced from the conversion of mass to energy is the binding energy. The energy to mass relationship is $E = mc^2$. The energy is in joules when the mass is in kg and the speed of light (c) is 3.00×10^8 meters /second.

15. The moderator in a nuclear reactor is needed to convert high energy neutrons to thermal neutrons. The thermal neutrons are required to start the fission of U-235 nuclei.

17. Three problems associated with nuclear power from fission are: catastrophic accident, plutonium production, disposal of radioactive wastes from the plant.

19. a. The "reprocessing of nuclear fuels" refers to the separation of useful plutonium and fissionable uranium from other nuclides in spent fuel elements.
    b. A number of dangers are linked to the reprocessing. The plutonium may be diverted for nuclear weapons. There maybe a breach of the process and radioactive material may escape into the environment. Plutonium itself is extremely toxic to humans.

21. Fission and fusion both rely on conversion of mass to energy. Fission requires splitting heavy nuclei into smaller ones while fusion involves joining small nuclei together to produce a larger nucleus.

	Fuels	Benefits	Problems	Current Status
Fission	Uranium Plutonium	renewable from breeder reactors	waste storage and disposal Pu toxicity	commercially available, proven method
Fusion	Deuterium Tritium	unlimited fuel supply	radioactive hardware from power plants, no containment vessel available	research only

23. a. The half-life of $^{239}_{94}Pu$ is 24,000 years.
    b. Plutonium is a problem because it does not dissipate quickly.

25. The "magnetic bottle" uses magnetic fields produced by strong electromagnets to contain a plasma by interacting with the magnetic fields generated by the ions in the plasma. This keeps the high temperature plasma from physical touching a container and dissipating the energy stored in the plasma.

27. The warming of the atmosphere and the earth's surface (water and land) is the process that is largest user of solar energy.

29. b. The annual energy use in the United States is approximately 84 quads.

31. Solar energy can be used to produce high temperatures if sunlight is concentrated using parabolic mirror concentrators. These can focus light on a small area, greatly increasing the amount of energy per square inch.

33. Geothermal energy is derived from the molten core of the Earth.

35. Geothermal and solar energy are abundant, but the conversion to usable forms of energy requires special devices. Some of these are extremely simple such as the solar cooker illustrated on page 401 in Figure 13.13. Some conversion devices are sophisticated like the Solar One electric generator system that uses computer controlled mirrors to focus sunlight on a boiler unit. The relatively simple photovoltaic cells require specialized manufacturing methods and materials that are not cheap. When these devices are produced on a large scale, their cost will decrease, but they are not "free".

37. Bioengineering may be used to develop plants that grow more rapidly. This would offer the possibility of more growing cycles in a year. Plants could be developed that require less intensive care and would be cheaper to grow. Another possibility is that plants could be developed that contain compounds that yield much higher energy per kilogram of plant material.

39. Each person will answer differently depending on the life style and energy sources used in the region where person lives.

Solutions to Selected Problems

1. a. The definition of a quad is 1×10^{15} Btu. The conversion method leads to a set up

$$8 \text{ quad} \times \frac{1 \times 10^{15} \text{ Btu}}{\text{quad}} = 8 \times 10^{15} \text{ Btu}$$

   b. The energy equivalent between (Btu) British thermal units and cubic feet of natural gas is 1 cubic foot natural gas = 1000 Btu;

$$1 \times 10^{9} \text{ cubic feet natural gas} \times \frac{1000 \text{ Btu}}{1 \text{ cubic foot natural gas}} = 1 \times 10^{12} \text{ Btu}$$

   c. The conversion of eight million British thermal units, 8×10^{6} Btu, to kcal requires the equivalency 252 kcal = 1000 Btu.

$$8 \times 10^{6} \text{ Btu} \times \frac{252 \text{ kcal}}{1000 \text{ Btu}} = 20.2 \text{ kcal}$$

3. The number of barrels of oil equivalent to 1 cord of wood can be calculated as follows:

$$1 \text{ cord firewood} \times \frac{28.8 \times 10^{6} \text{ Btu}}{1 \text{ cord}} \times \frac{1 \text{ quad}}{1 \times 10^{15} \text{ Btu}} \times \frac{180 \times 10^{6} \text{ barrels of oil}}{1 \text{ quad}} = \frac{5.2 \text{ barrels}}{1 \text{ cord}}$$

6. The energy released when one ton of coal is burned in one second is 26.4×10^9 J. The power released would be this energy divided by the time. Power (in Watts) is energy in Joules divided by time in seconds.

$$\frac{26.4 \times 10^9 \text{ joules}}{1} \times \frac{1}{1 \text{ second}} = 2.64 \times 10^{10} \frac{\text{joules}}{\text{second}} = 2.64 \times 10^{10} \text{ Watts}$$

8. The number of barrels of oil needed to heat a building with 500 offices can be calculated using the number of Btus required per office and the energy equivalent of a barrel of oil.

    a. Each day the energy requirement is $\dfrac{150 \text{ Btu}}{1 \text{ office}} \times \dfrac{500 \text{ offices}}{\text{building}} = \dfrac{75000 \text{ Btu}}{\text{building}}$;

    The number of barrels of oil is $\dfrac{75000 \text{ Btu}}{\text{building}} \times \dfrac{1 \text{ barrel}}{5.6 \times 10^6 \text{ Btu}} = \dfrac{1.3 \times 10^{-2} \text{ barrel}}{1 \text{ building}}$

    b. If the efficiency is only 30% when electric heat is used, then the number of barrels of oil needed would be greater than the 1.3×10^{-2} barrels used if the oil was burned in an oil furnace. If you let x equal the actual number of barrels required for the 30% efficient electricity production, then 4.3×10^{-2} barrels would be required to heat the building by electric heat.

1.3×10^{-2} barrels $= 0.3 (x)$

Solve for x $= \dfrac{1.3 \times 10^{-2} \text{ barrels}}{0.3}$

x $= 4.3 \times 10^{-2}$ barrels

Speaking of Chemistry

Name _____

Alternate Energy Sources

Across

1 Solar heating that captures sun's heat without pumps or fans.

6 Traditional biomass fuels like wood.

7 Country with the highest share of electricity produced by nuclear fission.

9 Power unit equal to 1 joule/ second.

11 A country with the lowest share of electricity produced by nuclear fission.

13 SI energy unit equaling $1 kgm^2/sec^2$.

15 Process in which one fission reaction provides particles for successive ones.

17 _____ energy determines the stability of a nucleus.

19 Graphite or water in a reactor that slows high energy neutrons without absorbing them.

Down

1 A state of matter consisting of unbound nuclei and electrons.

2 ____ energy delivers approximately 46000 quads to the U.S. each year.

3 Power unit equal to 1000 joules/sec.

4 Measure of the rate of energy consumption, energy per unit time.

5 Woman physicist who was a codiscoverer of nuclear fission in 1918.

8 Process of forming larger nuclei by combining light nuclei like H or Li.

10 Energy from the high temperature molten core of the Earth.

12 Radioactive product from U-239 with a half-life of 24,000 years.

14 _____ fuels like biodiesel fuel from soybean oil.

16 Natural U-238 must be _____ with uranium-235 to make it fissionable.

18 Amount of energy needed to raise the temperature of one pound of water by 1^O F.

Bridging the Gap I
Concept Report Sheet
Fuel Taxes and Oil Conservation

Name _____

Global petroleum reserves are a valuable resource providing fuel and starting compounds for manufacturing chemicals such as pharmaceuticals. Present use of oil as an energy source is under attack because the resource is limited and burning oil eliminates the possibility of using petroleum for other purposes. There are proposals aimed at reducing consumption in the United States by raising taxes on gasoline produced from oil (petroleum). The proposed new taxes would increase gasoline prices to a range of approximately $3.00 to $5.00 per gallon. This is the price range for gasoline in Western Europe and Japan. The presumption is that the increased price would push down demand and consumption of petroleum.

Your task is to do the following

1. Develop and record a question you would want answered by experts before you would decide to oppose or support the proposed taxes.

2. Answer the following questions
   What do you think this tax increase would do to the cost of manufacturing and shipping consumer goods and farm products?

   Do you think the increased fuel costs would influence demands for wage increases and inflation? How?

Do you think the increased petroleum gasoline prices would have an effect on the production and sale of automobiles? How?

What effect do you see on airline fares and other forms of public transportation if the price of fuel tripled? Do you think bus fuel and airline jet fuel should be exempted from this tax? Would you recommend subsidizing the airline and transportation industry?

200

Bridging the Gap II
Concept Report Sheet
Biomass Fuels and Food Production

Name_____

The production of biomass as a source of fuel for energy production is a real public policy option. At the present the farmers of the United States are producing surpluses of grain. This excess production is available for export and for production of alcohol. Projections of population growth show that the future population of China alone will increase to the point where the Chinese themselves could consume all of the grain production of Canada and the United States.

Assume you are a member of Congress. You are on a committee that is considering legislation which would require a gradual expansion of biomass fuel production and increased consumption by consumers. What three questions would you want your staff to research before you formed an opinion on the legislation?

Solution to Speaking of Chemistry crossword puzzle

Alternate Energy Sources

14 Organic Chemicals and Polymers

14.1 Organic Chemicals

This section describes the sources of organic chemicals. It tabulates the classes of compounds produced by distillation of coal tar. The historic discovery by Wöhler is described. Substitution and addition reactions are illustrated. The concept of functional groups is introduced and example compounds are shown. The standard suffixes used for naming alcohols, carboxylic acids and aldehydes are introduced.

Objectives

After studying this section a student should be able to:

tell how coal gas is produced
identify the process used to make ethylene, propylene, etc. from petroleum
explain what the "vital force" myth was and tell how Wöhler destroyed this myth
tell what a substitution reaction is and give an example
tell what an addition reaction is and give an example
tell what the term functional group means
identify a class of compounds from a formula

Key terms

organic chemicals	aromatic compounds	vital force myth
substitution reactions	hydrocarbons	synthesis
alkanes	addition reactions	alkenes
alkynes	hydrogenation	functional groups
alcohol, "-ol"	aldehyde, "-al"	carboxylic acid, "-oic"

14.2 Alcohols and Their Properties

Primary, secondary, and tertiary alcohols are defined and some important alcohols are described. Typical reactions of alcohols are illustrated such as oxidation to aldehydes and ketones, dehydration to alkenes, formation of ethers. Hydrogen bonding in alcohols is detailed. Alcohol toxicity, alcohol proof, and denatured alcohol are discussed. Ethylene glycol and glycerol are described.

Objectives

After studying this section a student should be able to:

tell how primary, secondary and tertiary alcohols differ
describe reactions of alcohols such as oxidation and dehydration
tell how hydrogen bonding in alcohols arises
tell how hydrogen bonding differs in alcohols with one, two, and three -OH groups
explain how proof and percent alcohol by volume are related to each other
tell how alcohol is denatured
describe the difference between ethanol and ethylene glycol

Key terms

primary alcohol	secondary alcohol	tertiary alcohol
"R- group" ·	R- and R´-	structural isomers
oxidation of alcohol	dehydration	hydrogen bonding
solubility	ethanol	fermentation

"proof"   ethanol metabolized   denatured
detoxification   acetaldehyde   acetic acid
ethylene glycol   permanent antifreeze   glycerol ("glycerin")
humectant   nitroglycerin

14.3   Functional Groups with the Carbonyl Group

This section describes the classes of compounds containing a carbon atom doubly bonded to an oxygen atom: aldehydes, ketones, carboxylic acids and esters. Common aldehydes and ketones are described. Simple carboxylic acids and naturally-occurring carboxylic acids are discussed. The preparation and properties of esters are also described.

Objectives

After studying this section a student should be able to:

> classify a compound as either an aldehyde, a ketone, a carboxcylic acid or an ester
> write the general formula for each of the following and circle the functional group:
> > aldehydes, ketones, carboxylic acids, esters
> tell what the natural sources are for citric acid, lactic acid, stearic acid, oleic acid
> tell what alcohol and acid are needed to prepare an ester given the name of the ester
> given the structural formula for an ester, tell what alcohol and acid were used to make it
> given an alcohol and an acid, draw the structural formula of their ester

Key terms

carbonyl group   ketone, "-one"   aldehyde, "-al"
carboxylic acid, ".-oic"   acetone   formaldehyde, methanal
-COOH functional group   formic acid   ester, "-ate"
acetic acid   weak acid   ethylacetate
dicarboxylic acid   food additives   flavors
hydrolysis

14.4   Organic Chemicals from Coal

This section describes how organic chemicals can be made from coal and steam. Specific reactions are illustrated for the production of synthesis gas, methanol, methyl acetate, and acetic anhydride.

Objectives

After studying this section a student should be able to:

> write the reaction for preparation of synthesis gas
> identify the structural formulas for methanol, methyl acetate, and acetic anhydride
> tell what products result from the reaction of water and acetic anhydride

Key terms

chemicals from coal   gasification of coal   synthesis gas
methanol   methyl acetate   acetic anhydride

14.5   Synthetic Organic Polymers

Definitions are given for plastics, monomers, polymers, addition polymers, and condensation polymers. Addition polymers such as polyethylene, and polymers derived from polyethylene like compounds are described. The structure and properties of natural and synthetic rubber are illustrated and discussed. Vulcanization of natural rubber is explained. Stereoregulation in synthetic rubber is described. The structure of neoprene rubber is illustrated.

Copolymers are defined and styrene-butadiene rubber, SBR, is illustrated. Condensation polymers such as polyester, polyamide (nylon), polycarbonates (Lexan™), phenol-formaldehyde (Bakelite™), silicon rubber, and Silly Putty are described.

Objectives

After studying this section a student should be able to:

give definitions for monomer and polymer

tell how addition polymers differ from condensation polymers

give definitions for free radical and initiation

tell how the polyethylene chain is formed from ethylene

give definitions for HDPE, LDPE, and CLPE

write the formula for styrene and for polystyrene

tell how styrofoam is made from polystyrene

identify the monomer that produces a specific polymer

identify monomers such as ethylene, vinyl chloride, styrene, etc. if given the structures,

tell how and why vulcanized rubber differs from natural rubber

describe a condensation reaction

identify the representative formula for nylon-66

identify the polycarbonate structure

identify a silicone polymer if given polymer structures

Key terms

polymer	monomers	addition polymers
initiation	free radical	ethylene, $CH_2 = CH_2$
polyethylene	HDPE	LDPE
CLPE	styrene	polystyrene
polypropylene	polyvinyl chloride, PVC	polyacrylonitrile
natural rubber	isoprene	latex
vulcanized rubber	elastomers	polybutadiene
poly-*cis*-isoprene	gutta-percha	*trans* polymer
stereoregulation	synthetic rubbers	neoprene
copolymers	Saran wrap	styrene-butadiene, SBR
condensation reaction	polyesters	PET
primary amine	nylon	amide
polyamides	amide linkage	nylon-66
Kevlar	polycarbonates	Lexan
condensation polymers	thermosetting	"cure"
silicone oils	organosilanes	silicone rubber

14.6   New Polymer Materials

This section describes reinforced plastics. The properties of reinforced plastics are compared with the properties of metals. Glass fiber and carbon fiber composites are discussed. Applications of composite materials are described.

Objectives

After studying this section a student should be able to:

give a definition for composites and reinforced plastics

describe how the strength and density of composites compare with those of aluminum and
steel

give a definition for specific gravity

name the fiber most commonly used in composites
describe how graphite/polymer composites are constructed
name applications for glass fiber and graphite/polymer composites

Key terms and equations

reinforced plastics	composites	polymer matrix
low specific gravity	high resistance	chemical resistance
glass fibers	polyesters	graphite fibers
graphite-epoxy composites		

14.7   Recycling Plastics

This section describes the present level of plastics recycling and tells why metals such as lead are extensively recycled. The requirements for successful waste recycling are identified. The code system for plastic identification is tabulated. The most commonly recycled plastics, PET and HDPE are described.

Objectives

After studying this section a student should be able to:
describe the main reason why plastics recycling lags behind recycling of lead and iron
identify the four phases needed for successful recycling of any waste
describe how plastic labeling aids consumers in identification of plastics for recycling
identify the two most recycled plastics and name the recycled applications

Key terms and equations

recycling plastics	number-one waste	collection
sorting	reclamation	end-use
PET	HDPE	most commonly recycled plastic
resin	poly(ethylene terphthalate)	

Additional Readings

Aihara, Jun-ichi. "Why Aromatic Compounds are Stable" (includes related articles and bibliography) Scientific American March 1992: 62.

Fisher, Harry L. "Rubber" Scientific American Nov. 1956: 74. I

"The Ultimate Science Story. (Molecular Modeling, Buckminsterfullerene Chemistry and AIDS in One News Story)" The Economist Aug. 21,. 1993: 69.

Mark, H.F. "The Nature of Polymeric Materials" Scientific American 197. 1967: 149.

Natta, Giulio. "How Giant Molecules Are Made" Scientific American Sept. 1957: 98.

Oster, Gerald. "Polyethylene" Scientific American Sept. 1957: 139.

Rochow, Eugene G. "The Chemistry Of Silicones" Scientific American Oct. 1948: 50.

Wingo, Walter. "Government Plans to Bolster 'Green' Design. (National Environmental Technology Strategy Promotes Environmentally-Friendly Technologies)" Design News June 26, 1995: 39.

Yam, Philip. "Plastics get Wired" Scientific American July 1995: 82.

Answers to Odd Numbered Questions for Review and Thought

1.   Carbon atoms can bond to other carbon atoms in almost unlimited numbers and in a variety of ways. Introducing other elements allows different molecules due to different atom sequences (functional groups and isomers).

3.

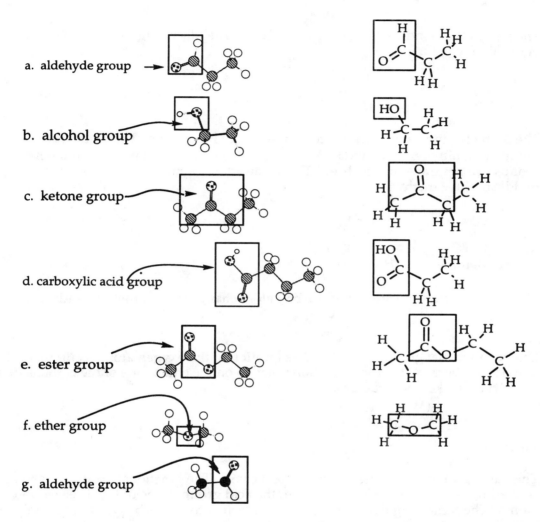

a. aldehyde group

b. alcohol group

c. ketone group

d. carboxylic acid group

e. ester group

f. ether group

g. aldehyde group

5.   a.   The -ol ending indicates the presence of an OH alcohol group. The name butan- indicates four carbon chain. The number 2 locates the alcohol OH group on the number two carbon.

a. 2-butanol

b.   Dimethyl ether is a symmetric ether. The methyl group has the formula CH_3-. The two methyl groups will be connected by an oxygen atom.

$$H-\underset{\underset{H}{|}}{\overset{\overset{H}{|}}{C}}-O-\underset{\underset{H}{|}}{\overset{\overset{H}{|}}{C}}-H$$

dimethyl ether

c.   The name ethan- indicates a two carbon chain. The -al ending indicates the presence of the aldehyde functional group.

$$H-\underset{\underset{H}{|}}{\overset{\overset{H}{|}}{C}}-\overset{\overset{O}{\parallel}}{C}\diagdown H$$

ethanal or acetaldehyde

d.   The prefix tri- indicates there are three methyl, CH_3-, groups. The amine functional group has a nitrogen atom with alkyl groups attached. The maximum number of alkyl groups in a neutral amine is three. This means the structure has a nitrogen in the middle of three methyls.

$$H-\overset{\overset{H}{|}}{\underset{\underset{H}{|}}{C}}-N-\overset{\overset{H}{|}}{\underset{\underset{H}{|}}{C}}-H$$

trimethylamine

e.   The butan- name indicates a four carbon atom chain. The ending -one indicates the

$$C-\overset{\overset{O}{\parallel}}{\underset{\underset{C}{|}}{C}}$$

ketone functional group,   . The location of the oxygen atom is automatically determined because the fourth carbon must go on the end of one of the carbons that are part of the ketone structure.

$$H-\overset{\overset{H}{|}}{\underset{\underset{H}{|}}{C}}-\overset{\overset{O}{\parallel}}{C}-\overset{\overset{H}{|}}{\underset{\underset{H}{|}}{C}}-\overset{\overset{H}{|}}{\underset{\underset{H}{|}}{C}}-H$$

butanone

f.   The name propan- indicates a three carbon chain. The functional group carboxylic acid with the structure COOH. The C in the acid group is one of the three carbon atoms. The remaining two carbons can be attached to the carboxyl group in only one way.

$$H-\overset{\overset{H}{|}}{\underset{\underset{H}{|}}{C}}-\overset{\overset{H}{|}}{\underset{\underset{H}{|}}{C}}-\overset{\overset{O}{\parallel}}{C}\diagdown O-H$$

propanoic acid

g. The name ethan- indicates a two carbon chain. The functional group name is a carboxylic acid with the structure,

$$-\overset{\overset{O}{\parallel}}{C}\diagdown OH$$ ethanoic acid

$$H-\overset{\overset{H}{|}}{\underset{\underset{H}{|}}{C}}-\overset{\overset{O}{\parallel}}{C}\diagdown O-H$$

208

h. The ethan- name indicates a two carbon chain. The -ol ending indicates the presence of an OH alcohol group. The di- prefix indicates two OH groups. The numbers 1 and 2 locate the two OH groups. There is one OH on each carbon atom.

$$
\begin{array}{c}
\text{H} \quad \text{H} \\
\text{H} - \text{C} - \text{C} - \text{H} \\
\text{1,2-ethanediol} \quad \text{H} - \text{O} \quad \text{O} - \text{H}
\end{array}
$$

7. There are four structural isomers for the alcohols $C_4H_{10}O$.

8.

9. a. pentanoic acid   b. 2-pentene
   c. diethyl ether   d. 3-pentanol
   e. 2-methyl-2-propanol   f. acetaldehyde or ethanal

11. a. Proof = 2 x percent alcohol by volume This means an 80 proof liquor is really 40% alcohol by volume and 60% water.
    b. Denatured alcohol is ethanol that has been adulterated with compounds that are toxic. These additives make the ethanol unfit for human consumption.

13. Pure ethanol has a proof rating of 200. It is 100 % pure x 2 = 200 proof.

15. An hydroxyl functional group, OH, is often present in naturally occurring carboxylic acids. An example is lactic acid which is present in milk.

lactic acid

$$
\begin{array}{c}
\text{H} \quad \text{H} \quad \text{O} \\
\text{H} - \text{C} - \text{C} - \text{C} - \text{O} - \text{H} \\
\text{H} \quad \boxed{\text{OH}}
\end{array}
$$
← carboxylic acid functional group
← alcohol functional group

17. a. Two naturally occurring esters are isoamyl acetate in bananas and butyl butanoate in pineapple. See Table 14.9 for some naturally occurring esters.
    b. Two naturally occurring carboxylic acids are formic acid in ants and lactic acid in milk. See Tables 14.7, 14.8 and 14.9 for simple carboxylic acids and naturally occurring acids.

19. The Eastman Kodak process to produce acetic anhydride from coal and water (steam) has the following four steps.
    $C + H_2O \longrightarrow CO + H_2$
    $CO + 2 H_2 \longrightarrow CH_3OH$
    $CH_3OH + CH_3COOH \longrightarrow CH_3COOCH_3$
    $CH_3COOCH_3 + CO \longrightarrow (CH_3CO)_2O$

21. The word polymer comes from the Greek words *poly* meaning "many" and *meros* meaning "parts". The literal meaning of polymer is "many parts".

23. a. A monomer is small molecule that can be linked to itself to make a larger molecule called a polymer

   b. A polymer is a large molecule made up of many small molecules (monomers) linked together

25. a. Polyethylene is a polymer built from CH_2CH_2 monomer units. The formula for polyethylene with "n" monomer units is shown for a typical structural formula

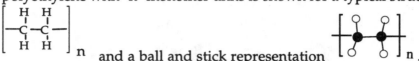 and a ball and stick representation 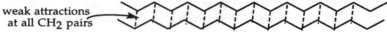 .

   b. HDPE refers to high density polyethylene which consists of long linear polyethylene polymer molecules that have a minimum of branching. The molecules can pack together well because they fit well against one another. The small attractive forces arise between a pair of CH_2 units on neighboring strands all along the chains.

weak attractions at all CH_2 pairs

   c. LDPE refers to low density polyethylene which consists of polyethylene chains that have multiple points where side chains branch off. The molecules do not pack as well so the density is lower than in HDPE. There are fewer groups that are attracted to one

weak attractions betweeen limited number of CH_2 pairs

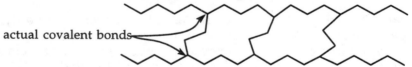

   another.

   d. CLPE refers to cross linked polyethylene which has polyethylene chains that are branched and cross linked to one another by the branches. This is a tougher polymer because there are actual covalent bonds between strands rather than just small induced attractive forces between the CH_2 units.

actual covalent bonds

27. The difference is in the positions of the CH_2 groups relative to the double bond: *trans* = across, *cis* = same side.

Poly-*trans*-isoprene, the CH_2 groups are on opposite sides of the chain.

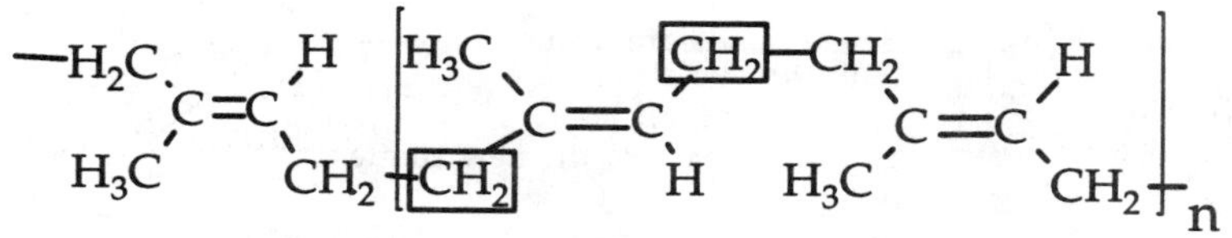

The poly-*trans*-isoprene is brittle and hard.

Poly-*cis*-isoprene, the CH_2 groups are on the same side.

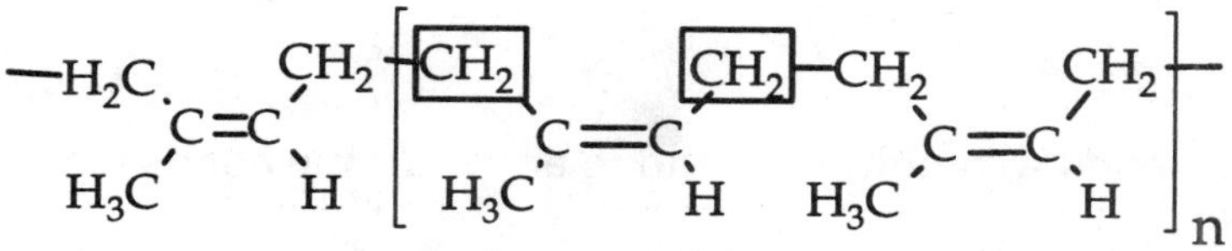

Poly-*cis*-isoprene is elastic and water-repellent.

29. It makes the rubber elastic and water-repellent but not sticky.

31. See also question 25 for more on this topic.

   a. In a linear unbranched polyethylene, The $-CH_2CH_2-$ unit repeats in a long line.

   b. In a branched polyethylene chain, there are side chains of $-CH_2CH_2-$ units attached to the main polymer chain.

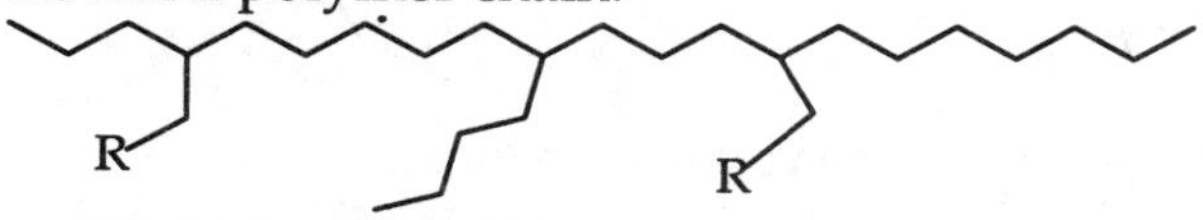

   c. a cross linked polyethylene

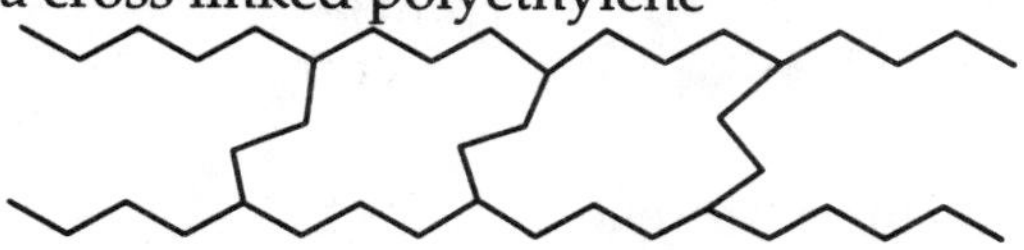

33.

   a. vinyl chloride

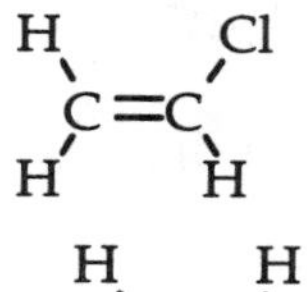

   b. styrene

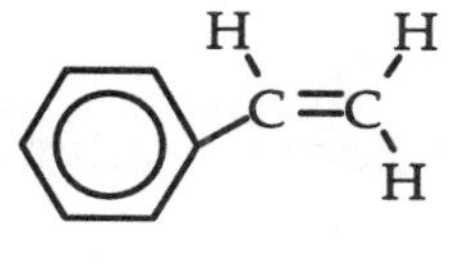

   c. butadiene

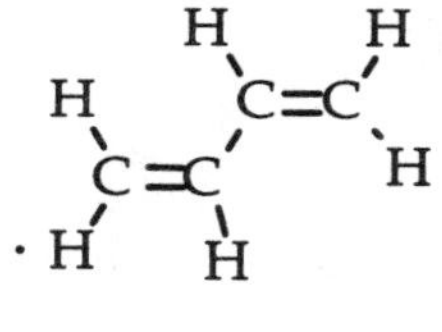

   d. propylene

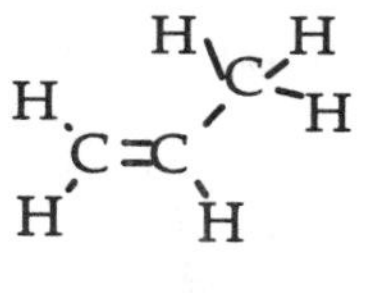

35. a. Yes, styrene can undergo an addition reaction because it can add to double bond in the ethene group. The addition converts the double bond into a single bond. An atom or group adds to both ends of the double bond. This process is illustrated here.

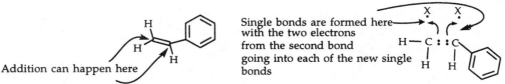

Addition can happen here

Single bonds are formed here with the two electrons from the second bond going into each of the new single bonds

b. Yes, propene can undergo an addition reaction because it can add to the double bond.

Addition can happen here

c. No, there are only single bonds so nothing can add to any multiple bonds in ethane.

37. a. Polyester is a polymer built from ester monomers.
b. Polyamide is a polymer built from amide monomers.
c. Nylon 66 is a polymer built from 6-carbon monomers. The condensation of hexamethylene diamine and adipic acid yields a polymer with 6 carbon atoms between the nitrogen atoms.

d. A diacid is a molecule containing two carboxylic acid groups.

e. A diamine is a molecule containing two amine groups.

f. A peptide linkage has a carbonyl group, $\diagup^{C=O}$, adjacent to an amine -NHR. group.

peptide linkage

39.

An organosilane is a compound with silicon bonded to alkyl groups as shown here.

41. Condensation reactions occur between two molecules to produce a larger molecule from the smaller reactant molecules and a separate small molecule like water which is eliminated.

43. Petroleum. Yes. Plant material sources are relatively cheaper so they will probably become the major source eventually. In addition the earth's petroleum resrves are limited and will not last indefinitely.

45. The four parts to successful recycling are: collection, sorting, reclamation, end-use. Unfortunately, collected materials will not be sorted or reclaimed if the supporting infrastructure for processing doesn't exist.

47. Composites give greater strength and they can be designed for a particular purpose. Yes, separation of fibers during reprocessing may be difficult.

49. PET in carpet fibers, HDPE in fencing and drain pipe

51. a. aldehyde, alcohol
    b. alcohol, aromatic, amine
    c. ketone, alkene, alcohol
    d. aromatic, aldehyde, ether, alcohol
    e. alcohol, carboxylic acid

Solutions for Selected Problems

1. The number of monomer units in a polymer is determined by dividing the molar mass of the polyethylene polymer by the molar mass of the ethylene monomer. Determine the molar mass for ethylene. One mol CH_2CH_2 = 28 grams CH_2CH_2.

$$\frac{280,000 \text{ grams}}{1 \text{ mole polymer}} \times \frac{1 \text{ mole ethylene monomer}}{28 \text{ grams}} = \frac{100,000 \text{ ethylene monomers}}{1 \text{ polyethylene polymer}}$$

4. a. Ethylene glycol and terephthalic acid form a polyester polymer and water.

ethylene glycol   terephthalic acid   poly(ethylene terphthalate)

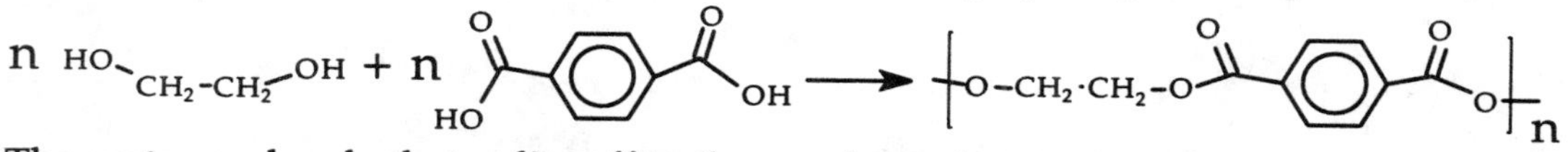

The water molecule that splits off in the condensation reaction is not shown.

   b. Ethylene forms polyethylene in an addition polymerization reaction.

ethylene monomers   polyethylene

Speaking of Chemistry

Name _____

Organic Chemicals & Polymers

Across

2 Natural rubber is too soft and sticky if it is not _____.
5 Small molecule used to make polymers.
6 Compounds like $CH_3CH_2OCH_2CH_3$, ROR.
7 Abbreviation for polyethylene terephthalate.
9 Compounds like acetic acid with the general formula RCOOH.
15 _____ catalysts catalyze reactions that favor formation of one geometric isomer.
17 Polystyrene
18 Trinitrotoluene, used as an explosive.
19 Synthesized by Wohler to disprove the "vital force" theory.

Down

1 Odor produced by 3-methylbutyl acetate.
3 Low density polyethylene.

4 _____ rubber is a hydrocarbon with the formula C_5H_8 is called latex rubber.
8 Reinforced plastics that contain fibers embedded in a plastic matrix.

10 Aromatic hydrocarbon with formula, C_6H_6.
11 Codes are stamped on plastic containers to help classify plastics for _____.
12 Polycarbonate polymer used for compact disks and bullet proof windows.
13 Alkene used to make butadiene.
14 Distillation of coal tar produces a _____ oil fraction that boils below 200°C.
15 Styrene-butadiene rubber is also called __ _____ rubber.
16 Compounds like CH_3NH_2.

Bridging the Gap
Concept Report Sheet

Name _____

Plastic Recycling: Classification and Sorting

Plastic recycling is legislated in many jurisdictions. The successful recycling of plastics or any material requires the four step recycling process of collection, sorting, reclamation, and end-use. A major problem with recycling plastics is that each type of plastic needs to be kept separate from the others to ensure a material that has an end-use market. If waste plastics are mixed, the reprocesser will not be dealing with a predictable polymer. The properties of reprocessed polymers can only be guaranteed by sorting the types of plastics. Right now consumers are asked to separate plastics and recycle using the codes shown here. Some recylcing agencies even go so far as to say that if you cannot classify the material with certainty, you should not guess and possibly contaminate a batch of good recyclable materials. This exercise is aimed at giving you the practical experience of identifying plastic products you have in your surroundings. Your task is to look at the recycling codes stamped on plastic items in your home, apartment or even on the shelves of stores. You are to identify the plastic and note what initial product the polymer was used to make. You may have trouble finding some polymers, but you should try to identify objects for each polymer.

Plastic code and polymer name	Kind of plastic item (toy, bottle, tool, etc.)	Kind of plastic item (toy, bottle, tool, etc.)
1 PETE — polyethylene terephthalate		
2 HDPE — high density polyethylene		
3 V — polyvinyl chloride (PVC)		
4 LDPE — low density polyethylene		
5 PP — polypropylene		
6 PS — polystyrene		
7 OTHER — other resins and multilayered multi-material		

Is it reasonable that some polymers are more difficult to find? Why? Give your answers on back of this page.

217

Solution to Speaking of Chemistry crossword puzzle

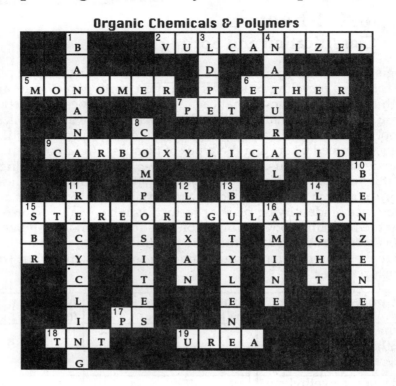

Organic Chemicals & Polymers

15 The Chemistry of Life

15.1 Handedness and Optical Isomerism

This section defines chiral, achiral, asymmetric carbons, and enantiomers. Examples of enantiomers are illustrated. Plane-polarized light, optical isomers and optical activity are described. The historic discoveries of Biot, van't Hoff, and Pasteur are described. Examples are given showing how the 2^n formula predicts the number of possible isomers for molecules containing "n" asymmetric carbons.

Objectives

After studying this section a student should be able to:

state the definition for chiral molecules and give examples
give the definitions for achiral molecules and enantiomers
inspect a structural formula and identify chiral atoms
describe how plane-polarized light interacts with chiral molecules
describe the discoveries of Biot, van't Hoff and Pasteur
apply the 2^n formula to predict the possible number of different isomers
give definitions for levo, L-isomers, and dextro, D-isomers
tell what type of chiral molecules are preferred in nature

Key terms

biochemistry	handedness	superimposable
nonsuperimposable	achiral	chiral
enantiomers	asymmetric carbon atom	tetrahedral
lactic acid	central carbon atom	dextro-
levo-	L-form	D-form
plane-polarized light	optical isomers	optically active
polarimeter	monochromatic light	stereochemistry
racemic mixture	2^n isomers	chiral atom
chiral drugs	thalidomide	aspartame, NutraSweet

15.2 Amino Acids

This section gives definitions for and examples of amino acids and zwitterions. Essential amino acids are defined. Formulas for 20 common amino acids are tabulated.

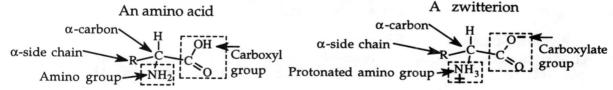

Objectives

After studying this section a student should be able to:

- draw and label the general formula for an amino acid
- tell how amino acids and proteins are related
- tell whether or not a molecule is an amino acid by looking at its structural formula
- explain what a zwitterion is and draw the structure for one if given the amino acid
- give the definition for essential amino acids
- tell the difference between amino acids that are basic, acidic, polar, and nonpolar

Key terms

condensation polymers	amino acids	L-enantiomer
amino group	carboxylic acid group	R group
alpha carbon	α-side chain	zwitterion
essential amino acids	lysine	tryptophan
methionine	nonpolar	polar
acidic	basic	

15.3   Peptides and Proteins

This section illustrates how a peptide bond is formed between an amino acid and an amine. The terms peptide, dipeptide, and polypeptide are explained. Simple proteins are defined and illustrated using insulin as an example. Conjugated proteins are defined.

Objectives

After studying this section a student should be able to:

- draw the general structure for a peptide bond
- give a definition for a peptide linkage
- give a definition for a dipeptide
- draw the peptide structure formed by a specific amino acid and an amine
- state the definition of an amino acid residue
- give a definition for simple proteins and name an example
- give a definition for conjugated proteins
- tell why proteins are called condensation polymers

Key terms

amide group	peptide	dipeptide
C-terminal end	"-ine" ending	-"yl" ending
amino acid residues	tripeptide	polypeptide
proteins	simple proteins	conjugated proteins
insulin	chymotrypsin	hemoglobin
myoglobin	condensation polymer	

15.4   Protein Structure and Function

This section shows how amino acid sequences can be indicated using abbreviations. The n! formula is explained using the 6 possible combinations for the tripeptides formed by Gly-Ala-Ser.

Gly-Ala-Ser	Ser-Gly-Ala	Ala-Gly-Ser
Gly-Ser-Ala	Ser-Ala-Gly	Ala-Ser-Gly

Neurotransmitters, enkephalins, and β-endorphins are described. The section defines the four levels of protein structure: primary, secondary, tertiary, and quaternay. Enzymes, active sites, and substrates are defined and enzyme action is explained.

Objectives

After studying this section a student should be able to:

give the definition for an amino acid sequence

tell the number of possible arrangements that can be formed by combining "n" amino acids

tell what the functions are for enkephalins and endorphins

give a definition for neurotransmitters

identify the amino acid sequence that enkephalins and β-endophins have in common, the amino acid sequence Try-Gly-Gly-Phe

use the amino acid sequence such as Ala-Gly-Ser using the structures in Table 15.1 to draw the tripeptide structure and give its name

tell what determines the primary structure of proteins

give a definition for the secondary structure for proteins

name the two most common secondary structures

explain what makes up the tertiary structure of a protein

tell what denaturation does to a protein, and tell what can denature a protein

describe the function of enzymes

give definitions for substrate and active site

explain what genetic diseases are

Key terms

amino acid sequence	number of arrangements	n!
natural opiates	enkephalins	endorphins
neuropeptides	receptor sites	β-endorphins
neurotransmitters	enzymes	hormones
intermolecular	intramolecular	primary structure
secondary structure	tertiary structure	quaternary structure
β-pleated sheet	disulfide bridge bonds	α-helix
left-handed helices	right-handed super helix	globular proteins
globular enzyme	denaturation	energy of activation
active sites	catalysts	substrate
hydroysis reaction	condensation reactions	reversible reaction
	genetic diseases	lactose intolerance

15.5   Carbohydrates

This section gives the general formula for carbohydrates, $C_x(H_2O)_y$. Definitions and examples are given for carbohydrates, monosaccharides, disaccharides, and polysaccharides. Example structures are given for α and β ring forms. Hydrolysis reactions of disaccharides are illustrated. Diabetes types I and II are defined. The role of artificial sweeteners (aspartame, saccharin, etc.) in treating diabetes and obesity is described. The structures and properties of polysaccharides, glycogen, cellulose, and starch are discussed.

Objectives

After studying this section a student should be able to:

give definitions for carbohydrates, monosaccharides, disaccharides, and polysaccharides

tell how disaccharides are formed from monosaccharides

describe how disaccharides are converted to monosaccharides

classify structures as mono-, di-, and poly- saccharides

give reasons for using artificial sweeteners

describe how Type I diabetes and Type II diabetes differ

tell how starches, glycogen, and cellulose differ

Key terms

carbohydrate	cellulose	monosaccharides
disaccharides	polysaccharides	hydrolysis
dynamic equilibrium	lactose	maltose
sucrose	$C_{11}H_{22}O_{11}$	invertase
invert sugar	aspartame	diabetes
saccharin	hypoglycemia	hyperglycemia
blood sugar	Type II diabetes	insulin
Type I diabetes	juvenile-onset diabetes	α-D-glucose chain
maturity-onset diabetes	glycogen	β-ring D-glucose
islets of Langerhans	amylopectin	dextrins
amylose		

15.6   Lipids

This section defines lipids. It describes waxes, fats, oils, and steroids. The general equation for the formation of fats (triglycrides) from fatty acids and glycerol is illustrated. Formulas for common fatty acids in oils and fats are given. Hydrogenation of vegetable oils is explained. The health concern about partially hydrogenated *cis* and *trans* fatty acids is discussed. Mono-unsaturated and polyunsaturated fatty acids are defined. Linoleic acid and prostaglandins are

discussed. The general structure for steroids is illustrated.   Cholesterol and sex hormones are described.

Objectives

After studying this section a student should be able to:

give definitions for lipids, waxes, fats, oils, steroids
give definitions for monounsaturated, saturated, and polyunsaturated fatty acids
if given a structural formula, classify the compound as a fat, fatty acid, steroid, or wax
tell how *cis* and *trans* unsaturated fatty acids differ
distinguish between a fat (ester) and a fatty acid
tell how a partially hydrogenated oil differs from an unsaturated oil
describe the function of prostaglandins
describe the problem that cholesterol poses

Key terms

fats	oils	triglycerides
triester	monounsaturated fats	essential fatty acid
prostaglandins	PGE_2	PGE_α
solid triglycerides	liquid triglycerides	hydrogenated
steroids	aromatic	four-ring skeletal structure
cis fatty acids	*trans* fatty acids	straight structures
cholesterol	alkyl group	alcohol group
sex hormones	progesterone	testosterone
estrogens	waxes	RCOOR'

15.7   Energy and Biochemical Systems

This section reviews photosynthesis and oxidation of glucose as energy sources. The structures for ATP and ADP are given. The roles of ATP and ADP in energy transfer are described.

Objectives

After studying this section a student should be able to:

describe the relationship between energy release and hydrolysis of ATP to form ADP
identify the structures for ATP and ADP
write the overall reaction for photosynthesis
tell how "light" reactions and "dark" reactions differ
explain why chlorophyll-containing plants are green (in terms of absorbed light)
write the overall reaction for oxidation of glucose
tell what is meant by "coupled reactions"

Key terms

adenosine triphosphate	ATP	adenosine diphosphate
ADP	photosynthesis	light reactions
dark reactions	chlorophyll	absorb violet and red light
transmit green light	excited electrons	coupled reactions

15.8   Nucleic Acids

This section gives the structures for the sugars,

 α-D-ribose and   α-2-deoxy-D-ribose.

The structures are illustrated for the five bases adenine(A), guanine(G), thymine(T), cytosine (C), and uracil(U) that make up nucleic acids. Nucleotides and trinucleotides are described. A trinucleotide sequence is shown. Complementary hydrogen bonding is discussed and illustrated. Protein synthesis is described in terms of mRNA, rRNA, and tRNA. Messenger RNA codons are tabulated for amino acids. Anticodons are defined and illustrated.

Objectives

After studying this section a student should be able to:

give definitions for nucleic acids, DNA, and RNA
name the five organic bases that form the message code in DNA and RNA
name the three organic bases found in both DNA and RNA
identify the structural components that are common to both RNA and DNA
tell how the structures for RNA and DNA differ
give definitions for nucleoside, nucleotide, and polynucleotides
give a definition for complementary hydrogen bonding
give definitions for gene and genome
describe the process of protein synthesis
if given a base sequence in DNA, tell what the base sequence is in complementary RNA
use a mRNA code table to tell what amino acids a codon series dictates

Key terms

deoxyribonucleic acid	nucleic acids	DNA and RNA
ribonucleic acid	organic bases	α-2-deoxy-D-ribose
α-D-ribose	adenine(A)	thymine(T)
cystosine(C)	guanine(G)	uracil(U)
ribosomal RNA	messenger RNA	transfer RNA
cytoplasm	nucleotides	nucleoside
polynucleotide	trinucleotide	complementary bases
genes	genome	complementary hydrogen bonding
template	replication	noncoding sequences
transcription	mitosis	meiosis
translation	codon	anticodon

15.9 Biogenetic Engineering

This section describes the process of using *E. coli* as gene factories. It illustrates the gene insertion process and gene recombination using tobacco mosaic virus and agrobacterium tumefaciens for altering tomato plant chromosomes. Examples of biogenetic engineering such as the production of transgenic animals and the biosynthesis of insulin are described.

Objectives

After studying this section a student should be able to:

describe the process of gene splicing
state reasons for and purpose of gene cloning
explain why bacteria are used to produce recombinant DNA
give reasons for developing transgenic species and identify two areas of research
give an example of a product from transgenic "animals"

Key terms

biogenetic engineering	recombinant DNA	transgenic
spliced gene	transgenic animals	human insulin
human growth hormone	tissue-plasminogen-activator	

15.10 Human Genome Project

This section describes the work of the National Institute of Health (NIH) to determine the complete sequence of base pairs in the human genome. The implications for diagnosing and treating hereditary diseases are discussed.

Objectives

After studying this section a student should be able to:

describe the Human Genome Project
tell why the NIH is carrying out the human genome project

Key terms

base pair sequence	automatic DNA sequencer	NIH
human genome	hereditary diseases	

Additional readings

Darnell Jr., James E. "RNA" Scientific American Oct. 1985: 68

Doolittle, Russell F. "Proteins" Scientific American Oct. 1985: 88.

Govindjee, Rajni. "The Absorption Of Light In Photosynthesis" Scientific American Dec. 1974: 68.

Hegstrom, Roger A. and Dilip K. Kondepudi. "The Handedness Of The Universe" Scientific American Jan. 1990: 108.

Kliener, Kurt. "Squabbling all the Way to the Genebank" (debate on accessibility of genetic information) New Scientist Nov. 26, 1994: 14.

McKenna, K. W., and V. Pantic, eds. "Hormonally Active Brain Peptides: Structure and Function" New York: Plenum Press, 1986.

Marshall, Eliot. "The Company that Genome Researchers Love to Hate (Human Genome Sciences Inc.)" Science Dec. 16, 1994: 1800.

Answers for Odd Numbered Questions for Review and Thought

1.   Chiral is derived from the Greek word "cheir" for "hand". It is used to describe objects that are non-superimposable mirror images of one another.

3.   A racemic mixture is a 50-50 mixture of enantiomers. The amount of optical activity produced by one enantiomer is canceled out by the activity of the mirror image.

5.   A zwitterion is a "double" ion that has both a + charge and a - charge.
      A neutral amino acid molecule is converted to a zwitterion by the shift of a proton, H^+, from the carboxylic acid to the amino group.

      The glycine zwitterion has a hydrogen for the R group.

      the R group is H

      glycine   zwitterion

7.   An essential amino acid is an amino acid that is not synthesized by the human body. An essential amino acid must be provided by the diet. Essential amino acids are valine, leucine, isoleucine, phenylalanine, methionine, threonine, tryptophan, and lysine.

9.   a.   A dipeptide is a molecule formed by linking two amino acids.
      b.   A polypeptide is a protein containing 3 or more amino acids.
      c.   An amino acid residue refers to the amino acids that are linked to form a peptide molecule.
      d.   A protein is a very large polypeptide containing between fifty to thousands of amino acid residues.

11. Neuropeptides are biomolecules that transmit chemical messages along nerve pathways by connecting with receptors. Examples: methionine-enkephalin, leucine-enkephalin.

13. a. starch   b. cellulose   f. proteins   g. DNA   h. RNA

15. Denaturing a protein changes its structure (this may be at any of the four structure levels, primary, secondary, tertiary, or quaternary) so it is unable to perform its normal function.

17. An active site is a part or region of an enzyme where the substrate sits down to interact with the enzyme.

Enzyme — Active site — HO — Substrate

19. a. A carbohydrate is a class of naturally-occurring polyhydroxy aldehydes or ketones.
    b. A simple sugar is a monosaccharide. It cannot be broken down into simpler, smaller sugar molecules by hydrolysis (reaction with aqueous acid).
    c. A disaccharide is a carbohydrate that can be broken down into two monosaccharides when hydrolyzed.
    d. A polysaccharide is a carbohydrate built of many monosaccharides bonded together.

21. A person with *juvenile-onset*, Type I diabetes, does not produce enough insulin due to damage to pancreatic cells that produce insulin. Insulin injections can help remedy this condition. A person with *adult-onset*, Type II diabetes, produces enough insulin, but the insulin receptors on cells do not function properly and fail to recognize insulin. This condition is treated by altering diet, increasing insulin levels to give functioning receptors more insulin to recognize and by taking oral medication.

23. A lipid is a naturally occurring organic substance in living systems that is soluble in nonpolar organic solvents, but insoluble in water.

25. Hydrogenation is the conversion of C=C double bonds to single bonds by adding hydrogen atoms to carbon atoms in the molecule.

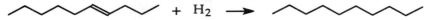

$+ \ H_2 \longrightarrow$

27. a. Photosynthesis is the process in which green plants absorb sunlight to make glucose and oxygen from water and carbon dioxide.
    b. Light reactions are the steps in photosynthesis that require light. The absorbed light excites the chlorophylls in the plant. Oxygen is produced from water and energy is stored in ATP.
    c. Dark reactions are the photosynthesis steps that occur in the absence of light. These steps use the energy stored in ATP to convert carbon dioxide and hydrogen from water into glucose and other carbohydrates.
    d. Chlorophyll is the molecule that absorbs light energy for use in photosynthesis.

29. DNA and RNA differ structurally because each contains a different sugar for the polymer chain and they contain different bases. DNA contains the sugar, α-2-deoxy-D-ribose,

$C_5H_{10}O_4$, 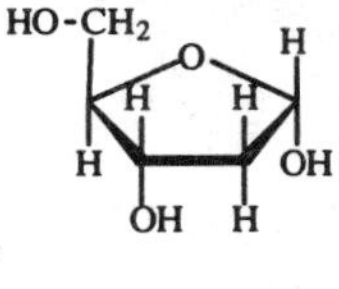 ,while RNA contains α-2-D-ribose, $C_5H_{10}O_5$, 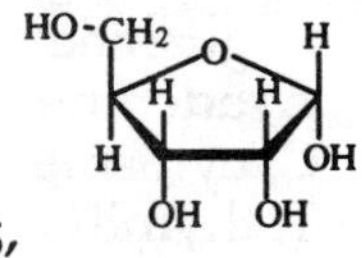 .

DNA contains the base thymine, 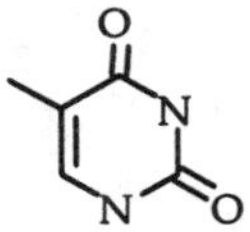 while RNA contains the base uracil.

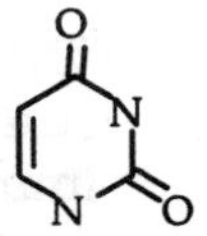

31. a. The number of combinations can be predicted using the formula, n!, where n equals the number of amino acids. This means that for a tetrapeptide there are, 4 x 3 x 2 = 24, possible combinations.

b. The 24 possible combinations for the amino acids glycine, Gly; alanine, Ala; serine, Ser: and cystine, Cys, are shown below. The combinations are determined by starting with one sequence and then systematically interchanging the positions of the amino acids. The table below was built up using this method.

Gly-Ala-Ser-Cys	Ala-Ser-Cys-Gly	Ser-Cys-Gly-Ala	Cys-Gly-Ala-Ser
Gly-Ala-Cys-Ser	Ala-Ser-Gly-Cys	Ser-Cys-Ala-Gly	Cys-Gly-Ser-Ala
Gly-Ser-Cys-Ala	Ala-Cys-Gly-Ser	Ser-Gly-Ala-Cys	Cys-Ala-Ser-Gly
Gly-Ser-Ala-Cys	Ala-Cys-Ser-Gly	Ser-Gly-Cys-Ala	Cys-Ala-Gly-Ser
Gly-Cys-Ala-Ser	Ala-Gly-Ser-Cys	Ser-Ala-Gly-Cys	Cys-Ser-Ala-Gly
Gly-Cys-Ser-Ala	Ala-Gly-Cys-Ser	Ser-Ala-Cys-Gly	Cys-Ser-Gly-Ala

33. The amino acids that are the in the tripeptide are valine, cysteine, tyrosine. The name for the tripeptide is Valylcystyltyrosine.

35. Enzymes are stereo specific. The D-glucose fits its active site, but the L-glucose will not act as a substrate for the enzyme.

37. The monomers that polymerize to form DNA and RNA are nucleotides. The nucleotide monomers contain a phosphoric acid unit, a ribose unit and a nitrogenous base. See Figure 15.25 on page 492.

39. Complementary bases are pairs of bases like adenine-thymine and guanine-cytosine. These pairs have two hydrogen bond sites between the A-T pair and three hydrogen bonds between the G-C pair.

41. The human genome project is intended to completely map the sequence of base pairs in human DNA.

3. The base pairs between DNA and mRNA are

DNA	mRNA
G, guanine	C, cytosine
A, adenine	U, uracil
C, cytosine	G, guanine
T, thymine	A, adenine

This means the best way to determine the sequence of bases in a complementary mRNA strand is to write out the DNA sequence and then one by one write the matching base immediately below.

a. The DNA strand with the base sequence T G T C A G T G G G C C G C T has a complementary mRNA sequence of A C A G U C A C C C G G C G A. The pairings look like this

DNA sequence   T G T C A G T G G G C C G C T
mRNA sequence   A C A G U C A C C C G G C G A.

The complementary pairings of bases between RNA strands are

mRNA	tRNA
G, guanine	C, cytosine
A, adenine	U, uracil

The tRNA anticodon order can be determined by writing out the mRNA sequence and then translating the complementary pairs one by one. There are essentially only two pairs to deal with so translation is not complicated.

b. Here the pairings would look like this

mRNA sequence   A C A G U C A C C C G G C G A

tRNA sequence   U G U C A G U G G G C C G C U

The anticodon order in the tRNA would be U G U C A G U G G G C C G C U

c. The mRNA codes for amino acids are tabulated in Table 15.6.
The table shows that the sequence of fifteen bases contains only five codons
U G U C A G U G G G C C G C U

[U G U] [C A G] [U G G] [G C C] [G C U] The amino acids that match these codons are:

U G U for Cysteine   C A G for Glutamine   U G G for Tryptophan
G C C for Alanine   G C U for Alanine.
The amino acid sequence is Cysteine Glutamine Tryptophan Alanine Alanine.

45. a. The structure for the amino acid with R = -CH₂-SH is

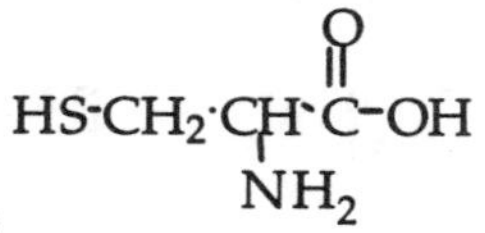

$$HS\text{-}CH_2\text{-}CH\text{-}\overset{\displaystyle O}{\overset{\|}{C}}\text{-}OH$$
$$NH_2$$

b. This amino acid is called Cysteine.

228

47. Watson and Crick proposed that the double helix was stabilized by hydrogen bonding between specific bases that had complementary hydrogen bonding. In these cases bases are attracted more to one another because of the match between sites in both members.

49. Two examples of successful applications of recombinant DNA technology are insulin production by bacteria, TPA production from transgenic animals

51. The amount of artificial sweetener needed is extremely small and distributing the sweetener uniformly through the product is difficult.

53. a. Guanine, Uracil, Cytosine are present in both mRNA and DNA.
    b. Adenine is the amino acid that is the fourth base found in mRNA.
    c. Threonine is the fourth base found in DNA but not in mRNA.

55. There are no wrong answers; each individual will have their own opinion.

Speaking of Chemistry

Name _____

The Chemistry of Life

Across

4 Geometric arrangement where groups are on opposite sides of molecular structure.

6 Discovered optical activity in 1811.
9 Stop codon.
11 Process of forming mRNA from a section of unwound DNA.
14 Carbohydrates like sucrose or maltose.
18 Tissue-plasminogen-activator used to dissolve blood clots in heart attack victims.
19 Liquid triglycerides.
20 Chiral molecules are _____ isomers.
21 Structure with groups on the same side of structure such as $CH_3CHCHCH_3$.

22 Female sex hormone containing two alcohol functional groups.

23 Bonds between atoms that share electrons unequally.

Down

1 Sequence of amino acids in structure of proteins.
2 Ribonucleic acid.
3 Three base code.
5 One of the essential amino acids.
7 Codon for start or methionine.
8 Amino acids form this double ion.
10 Adenine triphosphate.
12 Solid triglycerides.
13 Start codon.
14 Simple sugar also known as blood sugar.
15 Adenine diphosphate
16 Protein produced by genetically engineered E.coli.
17 Separated enantiomers of tartaric acid in 1848.

Bridging the Gap   I
Tetrahedron Construction and Chiral Centers

Two dimensional illustrations in the chapter are excellent, but they cannot give a tactile sense of the tetrahedral form and the non superimposability of molecules that are mirror images of one another. Enantiomers are molecules that have the same molecular formula but different spatial arrangements. They are mirror images of one another. This exercise is intended to give you an experience with three dimensional models of the lactic acid enantiomers.

Please read all these directions before doing any cutting.

Write your name in the blank space provided on the templates. Your instructor may want you to turn in your completed tetrahedra. Be careful to keep the A, B, and D tabs on the template when you cut out the tetrahedron. Be sure to leave the black edges on the faces. Color code the four different groups at the corners of the tetrahedron before cutting. A suggested color code is given here.

Group	hydrogen, -H	hydroxyl, -OH	methyl, -CH₃	carboxylic acid, -COOH
Color	white	pink	green	red

Hold the cutout so you can read your name. Fold faces A, B, and D away from you. Hold face D up so you can read it. Fold the hidden support away from you. Do this same process with face B and the second hidden support. Slide the hidden support behind face A. Fold tab B over the B on face B. Insert the remaining hidden support behind face B. Fold tab A over the A on face A. Fold tab D over the D on face D. All the tabs can be secured with a piece of transparent tape if you wish. You now have your tetrahedron that represents lactic acid.

The methyl group (-CH₃), hydroxyl (-OH), carboxylic acid group (-COOH), and hydrogen (-H) are on the corners of the tetrahedron. A carbon atom is in the center of the tetrahedron. There are two different tetrahedral structures for lactic acid, one is the mirror image of the other. After constructing the two tetrahedra hold a small mirror next to one of the models. You should rotate the other model to match the appearance of the image in the mirror. You now can see why the pairs are called mirror images of one another. Put away the mirror and try to rotate the two models to see if you can match the positions of the colors (substituents). If you can, they are superimposeable. If you cannot match them point for point at all points, then the models are nonsuperimposable.

The drawing below illustrates two isomers that are mirror images of one another.

233

Bridging the Gap
Templates for chiral tetrahedra

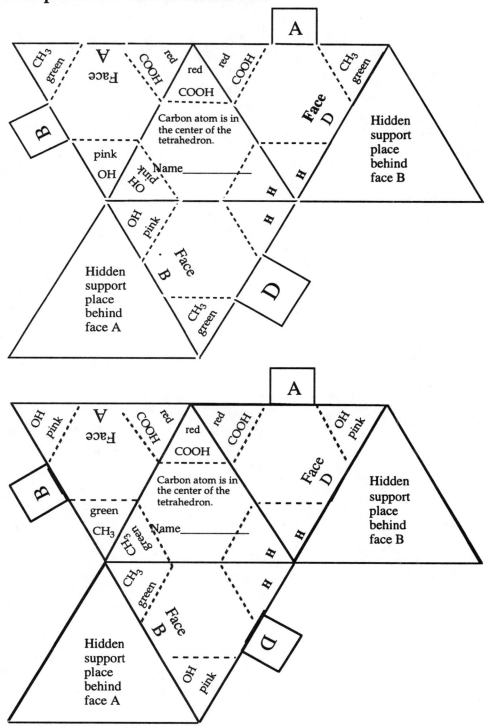

236

Solution to Speaking of Chemistry crossword puzzle

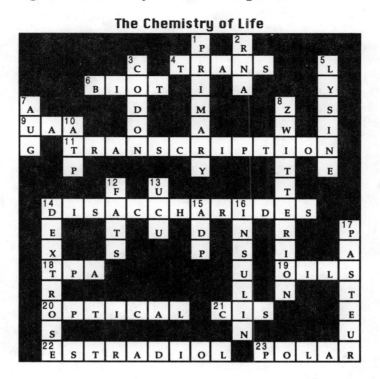

The Chemistry of Life

237

16 Consumer Chemistry--
Looking Good and Keeping Clean

16.1   Cosmetics
This section gives a definition for cosmetics and explains how drugs differ from cosmetics. The Food, Drug and Cosmetics Act and the Food and Drug Administration (FDA) are discussed. Trends in cosmetic development are also discussed.

Objectives
After studying this section a student should be able to:
give definition for cosmetics
give traditional distinction between drugs and cosmetics
tell what FDA does
describe the different standards used by the FDA

Key terms
cosmetics	drugs	FDA
aesthetic	ingredient listing	

16.2   The Chemistry of Skin and Hair
This section describes and illustrates the structure of the skin and hair. The dermis and epidermis layers of the skin are discussed. Chemical peels such as alpha-hydroxy acids (AHA) are defined and their action is explained. The function of the sebaceous glands is described. The nature of eccrine and apocrine sweat glands is explained. Hair keratin and the role of cysteine in hair protein is illustrated.

Objectives
After studying this section a student should be able to:
tell how the dermis and epidermis layers make up the skin
explain how chemical peels work
describe the action of the sebaceous glands in the dermis
describe the function of the sweat glands
tell how the eccrine sweat glands differ from the apocrine sweat glands
explain how the level of cysteine in hair influences the structure of a hair strand

Key terms
functions	protection	sensation
excretion	temperature control	stratum corneum
epidermis	keratin	chemical peels
trichloroacetic acid	glycolic acid	sebaceous glands
sebum	dry skin	excessively moist skin
sweat glands	eccrine glands	apocrine glands
cysteine	disulfide bridges	

16.3   Creams and Lotions

This section gives the reasons for treating dry skin with oily substances. Skin moisturizers and emollients are defined. The function of creams, lotions, emulsifying agents, and emulsions are explained. Colloids, oil-in-water emulsions, and water-in-oil emulsions are described. Cold cream is defined and a typical cold cream formulation is tabulated.

Objectives

After studying this section a student should be able to:

> give definitions for emollients and moisturizer
> name the two types of mixtures used as emollients
> give the definition for an emulsion
> explain the function of a barrier cream
> tell what a colloid is and give an example
> tell how to classify an emulsion as oil-in-water or water-in-oil
> explain how the cosmetic "cold cream" gets it name
> explain how the "dispersed phase" differs from the "continuous phase"

Key terms

dry skin	oils	moisturizer
creams lotions	emollients	emulsions
emulsifying agent	oil-in-water	water-in-oil
colloid	dispersed phase	continuous phase
barrier cream	shelf life	settle out
cold cream		

16.4   Lipsticks and Powders

This section gives a definition for lipsticks and powders. It explains the purpose of lipsticks and tells what types of compounds are used to make them. A typical lipstick formulation is given in Table 16.3. The compounds used to give color to lipsticks are described. The function of body powders is explained. A representative body powder formulation is listed in Table 16.4. Astringents are defined and their mode of action is described. The role of zinc oxide in astringents is explained.

Objectives

After studying this section a student should be able to:

> tell what function lipsticks serve
> name the major components in lipsticks
> explain how lipstick colors are usually produced
> tell what powders do for the skin
> name the principle ingredients in body powders
> describe the action of zinc oxide in an astringent

Key terms

lipstick	coloring agents	waxes
perfumed	flavored	dyes
metal ions	precipitated	powders
talc	corn starch	calcium carbonate
binder	astringents	zinc oxide, ZnO

16.5 Sunscreen Products

This section reviews the nature of light. It describes the ultraviolet range of light, defining uvA and uvB. It describes how the skin responds to ultraviolet light to produce melanin. The minimum erythemal dose (MED) is defined. The relation between stratospheric ozone and uv levels is explained. The uv index developed by the National Oceanic and Atmospheric Administration (NOAA) is described. The purpose of sunscreens and their mode of action is explained. The sun protection factor (SPF) is defined. The role of the FDA in regulating SPF products is described. The cancer risk posed by tanning and ultraviolet exposure is explained.

Objectives

After studying this section a student should be able to:

tell how the energy carried by ultraviolet light compares with the energy of visible light
tell how uvA and uvB light differ
describe how the skin responds to exposure to uvA and uvB light
tell how melanin concentrations and skin color relate
give a definition for erythema
give the definition for MED
explain how stratospheric ozone concentrations are related to uv levels at ground level
tell how SPF values for sunscreens are determined
explain what sunscreens do to protect the skin from sunlight
tell why a uv index was developed by NOAA
describe what can cause precancerous skin lesions

Key terms

ultraviolet (uv) radiation	wavelength	uvB
uvA	nm, nanometer	melanin
erythema, skin irritation	minimum erthymal dose	MED
stratospheric ozone	sunscreens	light absorbing
PABA	sun protection factor, SPF	absorption of light
cumulative exposures	cumulative effects	precancerous skin lesions

16.6 The Chemistry of Perfumes

This section tells how perfumes are formulated to have odors that are identified as a top note, middle note, and end note. It traces the historical development of perfumes, ranging from those only made from natural essences to those like Chanel No. 5™ which contain synthetic chemicals. The primary odors detected by humans are described.

Objectives

After studying this section a student should be able to:

give the definitions for the three notes that are typically found in perfumes
tell what is meant when a compound is described as a volatile compound
name the primary odors detected by humans
explain what a receptor site does in detecting odors
name the first perfume to use synthetic ingredients
tell how odors relate to molecular shape and odor receptors
name the most abundant compound in many perfumes
tell how perfumes differ from colognes and after-shave lotions

16.7 Deodorants

This section explains that sweat is odorless but that body odors result from bacterial action on amines and other compounds in sweat. The three general types of deodorants are described: astringents, chemicals that mask odors, and compounds that remove odorous substances by reaction. Specific astringents like $AlCl_3 \cdot 6\ H_2O$ are discussed. The "trial and error" aspect of deodorant effectiveness is discussed.

Objectives

After studying this section a student should be able to:

explain why relatively odorless sweat can lead to body odor
tell what an astringent does
explain why antibacterials are often included in deodorants
name a compound that has astringent action
name the types of compounds that are used to mask body odors
tell what zinc peroxide (ZnO_2) does to control body odor
tell how the timing for using astringents and odor masks differ
name a possible advantage of deodorants that are clear

Key terms

16.8 The Difference Between a Good Hair Day and a Bad Hair Day

This section describes the ionic and hydrogen bonds that exist between hair protein strands. The effects of water and pH on the bonds in hair are described. The chemical changes that occur in the permanent wave process are described. The reactions and compounds involved in hair color changes are discussed. The three types of hair dyes (temporary, semipermanent and permanent) are defined. The formulas and structures for hair sprays and depilatories are illustrated.

Objectives

After studying this section a student should be able to:

tell why minoxidil is a popular product
describe the ionic bonds that exist between adjacent protein strands
describe the effect of water on ionic bonds between proteins
tell how hydrogen bonds are formed between adjacent protein strands
describe the effect of water on hydrogen bonds between protein strands
explain how low pH solutions affect hair structure
tell what the normal pH is for hair
explain what effect ammonium thioglycolic acid has on disulfide bonds in hair
tell why an oxidizer is needed in the permanent wave process
tell what relation exists between hair color and the amount of melanin in a hair
tell how temporary, semipermanent , and permanent hair dyes differ

describe how hair color is changed by hydrogen peroxide, H_2O_2
describe how hair sprays hold hair in place
give a definition for a depilatory
why do depilatories irritate skin
state the pH range for water soluble sulfides like Na_2S

Key terms

minoxidil	sulfide linkages	hydrogen bonds
ionic bonds	static electrical charge	pH
disulfide cross-links	permanent wave	reducing agent
neutralizer	oxidizing agent	melanin
temporary coloring	semipermanent dyes	permanent coloring
bleaching	permanent dyes	oxidation dyes
hair sprays	resins	copolymer
solvent	propellant	depilatory
water-soluble sulfides	electrolysis	

16.9   Keeping Things Clean--Including Ourselves

A brief history of soap starts with the Sumerians in 2500 B.C. Definitions for surfactants soaps and saponification are given. Example reactions illustrate the formation of a soap and glycerine from tristearin (an animal fat) and sodium hydroxide. The soap making process used by frontier farmers and pioneers is described. The cleaning action of soap is described in terms of the molecular structure of soap and the formation of an emulsion.

sodium stearate (a soap)

organic soluble nonpolar end

water soluble ionic end

The emulsifying action is discussed. The reaction between soap and "hard" water ions $(Ca^{2+}, Mg^{2+}$ and $Fe^{2+})$ is used to explain the formation of soap "scum" or the proverbial bath tub ring.

Objectives

After studying this section a student should be able to:
give the definition for surfactants
give the approximate year in recorded history for the discovery of soap
describe the saponification process for making soap
give a definition for "lye" , "soap" , " glycerin"
sketch a soap molecule and identify the hydrophilic part and the hydrophobic part
name the three metal ions that are found in "hard" water
explain how soap scum forms
explain how soap forms an emulsion when it dissolves oil

Key terms

surface tension	surfactants	soap
fat	alkali	saponification
salted out	hydrocarbon chain	nonpolar
hydrated	emulsion	hard water

16.10   Synthetic Detergents

This section describes the properties and structures of synthetic detergents. Anionic surfactants are defined and an example like the one below is illustrated. Similarly cationic surfactants are

described and a general formula is illustrated. Nonionic detergents are defined and a typical structure is shown. The specific properties of each type of surfactant are described. Applications for each type of surfactant are given.

Objectives

After studying this section a student should be able to:

tell how synthetic detergents differ from soap
give a definition for hydrophobic and for hydrophilic
sketch a detergent molecule and identify the hydrophobic and hydrophilic portions
identify the negatively charged groups in anionic detergents
sketch the general structure for an cationic detergent
explain why a long hydrocarbon chain is not attracted to water molecules
tell what happens when an anionic detergent mixes with a cationic detergent
explain why ionic groups are attracted to water molecules

Key terms

hydrophobic	hydrophilic	alcohol
oil soluble	water soluble	anionic surfactants
cationic surfactants	quaternary ammonium halides	fabric softeners
nonionic	ester linkage	ether linkage

16.11   Shampoos--Detergents that Clean our Hair

Shampoos are described in terms of the surfactants and compounds needed to provide the desired properties. The structure and attributes of nonionic, anionic, and cationic surfactants are discussed. Representative structures for each class are illustrated. The chemical reasons why rinses solve "fly away hair" problems are given. The purpose of shampoo ingredients like lanolin and oil is explained.

Objectives

After studying this section a student should be able to:

give a definition for a shampoo
name the type of detergent most frequently used to formulate a shampoo
tell which one of the three types of detergents generally is a better foaming agent
state the reason for adding nonionic surfactants to shampoos
tell what if any connection there is between lather and cleaning efficiency
tell why a cationic rinse is used after washing hair with an anionic shampoo
explain why a cationic rinse makes damaged or disrupted hair feel smooth
give two reasons for adding lanolin and mineral oils to shampoo

Key terms

shampoos	anionic detergent	surfactants
cationic detergent rinse	conditioner	moisture

16.12   Toothpaste--Detergent Mixtures We Put in our Mouths

Formulations of toothpaste are described. The two major compounds in tooth enamel are identified. The origins of plaque and tartar are described. Toothpaste ingredients that fight plaque, tartar and gum disease are identified. The reason for adding stannous fluoride to toothpaste is explained.

Objectives

After studying this section a student should be able to:

tell what two main types of substances are present in toothpaste
tell what role calcium carbonate and calcium hydroxy phosphate play in tooth enamel
describe what happens when acid contacts tooth enamel
describe how dental plaque and tartar form and tell how they differ
explain why sodium pyrophosphate, $Na_4P_2O_7$, is a toothpaste ingredient
name two abrasives and tell why they are in toothpaste
explain why a surfactant or detergent is an ingredient in toothpaste
explain how fluoride hardens tooth enamel
name two sources of fluoride ion in toothpaste
describe how tooth loss from gum disease and from tooth decay compare

Key terms

toothpaste	abrasive	plaque
tooth enamel	calcium carbonate	apatite
calcified plaque	tartar	$SiO_2 \bullet nH_2O$
hydrated silica	fluoride ion	fluoride in drinking water
stannous fluoride	gum disease	decay

16.13   Bleaches and Whiteners

A short summary of the history of bleaches and whiteners is presented. Definitions for bleaching agents and optical brighteners are given. The reactions of bleaching agents to remove stains are discussed. Claude Louis Berthollet's discovery of the bleaching action of chlorine is described. The fluorescence process that makes optical brighteners work is explained.

Objectives

After studying this section a student should be able to:

give the definition for bleaching agents
name two common bleaching agents
describe the process by which bleaching agents remove unwanted colors
describe Berthollet's role in the development of bleaching agents
give the definition for fluorescence
explain how optical brighteners function in terms of fluorescence
explain why " black light " makes a shirt washed in an optical brightener glow white

Key terms

bleaching agents	oxidizing agents	sodium hypochlorite
chlorine	oxygen	light-absorbing
optical brighteners	fluorescence	reflected light

16.14   Stain Removal

This section describes the action of stain removers. The stains and appropriate removers are tabulated in Table 16.6. Stain removal is explained in terms of reactions and solubility. The removal of oil and fat stains by tetrachloroethylene ($Cl_2C=CCl_2$) is discussed. The reaction of sodium thiosulfate ($Na_2S_2O_3$) to remove iodine stains is described in detail.

Objectives

After studying this section a student should be able to:

explain why tetrachloroethylene is a good solvent for oil and grease stains
tell why iodine stains are removed by thiosulfate
tell the appropriate stain remover for a specific stain such as lipstick

Key terms

solubility	oxidation	chlorinated solvents
tetrachloroethylene	$Cl_2C=CCl_2$	

16.15   Corrosive Cleaners

This section describes cleaners with extreme values in pH. The applications and properties of strong acidic cleaners like hydrochloric acid solutions (HCl) are discussed. The applications and properties of strong alkaline cleaners like sodium hydroxide (NaOH) are described. Safe practices in using these products are discussed.

Objectives

After studying this section a student should be able to:

tell the pH range for strong acid cleaners like HCl and H_3PO_4 solutions
describe the proper way to handle corrosive cleaners
tell what cleaning chores acid solutions are used for
name the most common strong alkali used in drain cleaners and oven cleaners
explain why drain cleaner often contains both aluminum hydroxide and sodium hydroxide
tell why aerosol spray oven cleaners must be used carefully

Key terms

corrosive cleaners	toilet-bowl cleaners	phosphoric acid, H_3PO_4 (aq)
pH = 2	hydrochloric acid, HCl	oxalic acid, $H_2C_2O_4$ (aq)
drain cleaners	pH = 12	sodium hydroxide, NaOH
hydrolysis	oven cleaners	caustics

Additional Readings

Boyer, Pamela. "Decoding Cosmetics Labels" <u>Prevention</u> Oct. 1995: 162.

"Can Soaps Do More Than Clean?" (liquid hand and bath soaps; includes article on body washes and shower gels) <u>Consumer Reports</u> Nov. 1995: 730.

Cannon, Clare. "The Fresh New Face of AHAs" (alpha hydroxy acids) <u>Harper's Bazaar</u> Mar. 1995: 106.

"Choosing a Lipstick" (Includes related articles on purchasing lipstick) <u>Consumer Reports</u>. July 1995: 454.

Clifford, Catherine. "How Good Will You Look in Ten Years?" <u>Redbook</u> Oct. 1995: 46.

Dombrink, Kathleen J. and David O. Tanis. "pH & Hair Shampoo" ChemMatters April 1983: 8.

George, Leslie. "Choosing the Right Toothpaste for You" Glamour Nov. 1993: 88.

Kushner. L. M. and Hoffman. J. "Synthetic Detergents" Scientific American Oct. 1951: 26.

Layman, Patricia. "Detergent Phosphates Revisited in Study." (Phosphate-free detergents are harmful to the environment) Chemical and Engineering News Feb. 7,1994: 13.

"Moisturizers" (Includes related articles) Consumer Reports. Sept. 1994: 577.

Rinzler, C. A. "The New and Improved Chemistry of Cosmetics" Science April 1982: 61.

Stinson, S. C. "Consumer Preferences Spur Innovation in Detergents" Chemical and Engineering News, Jan. 26, 1987: 21.

Wandycz, Katarzyna. "Can Sunscreens Kill You?" Forbes July 19, 1993: 212.

"What Your Skin Really Needs" Mademoiselle. Feb. 1996: 144.

Answers for Odd Numbered Questions for Review and Thought

1. Protein chains in hair are held together by hydrogen bonds, ionic bonds and disulfide bonds. Hydrogen bonds can form between H atoms that are attached to very electronegative atoms (N and O) and the electron rich atoms with lone pairs such as N and O. The ionic bonds result when carboxylic acids (RCOOH) lose a proton to form carboxylate ions (RCOO$^-$) and the basic amino groups (-NH$_2$) gain a proton to form -NH$_3^+$; these oppositely charged structures are attracted to one another. Disulfide bonds form between sulfur atoms in cysteine fragments in adjacent strands and the strands are held together by disulfide bonds, -S-S-, between cystine residues.

3. The disulfide bond is a single bond between two sulfur atoms. The disulfide bonds between cysteine amino acid units hold parallel strands of hair protein in place.

Cysteine unit · Disulfide bond

5. The choice between mink oil and a different animal oil should take into account the following: cost, compatibility with skin, effectiveness, the humaneness of the method used to obtain the oil, and the survivability of the species.

7. Ingredients are named in decreasing order of abundance in the formulation. The listing of water after the names of the oils indicates that oils are the dominant part of the mixture. The mixture is a water-in-oil emulsion.

9.	Tanning results from uv radiation striking the skin and increasing the concentration of melanin pigment. If uv exposure is too great, erythema (reddening of the skin) occurs.

11.	PABA absorbs uvB before it can reach skin surface. In contrast, zinc oxide tends to reflect light and not absorb it.

13.	The top note is most volatile, middle note is intermediate, and end note is the least volatile and therefore, lasts the longest.

15.	Colognes are diluted perfumes, about one-tenth as strong as perfumes. Perfumes are 10%-25% perfume essence and 75-90% alcohol. A fixative is also added to keep the fragrant oils dissolved. Colognes have about 1%-2.5% perfume essence and about 98% alcohol.

17.	Disulfide bonds in the hair are broken by a reducing agent. The hair strand is shaped, and then the disulfide bonds are reformed using an oxidizing agent.

19.	Hair sprays are mixtures of a volatile solvent and a resin. The resin is the residue left behind on the strand of hair. The resin is supposed to hold the hair strand in place.

21.	Hair color products contain *para*-phenylenediamine. *Para*-phenylenediamine is used to produce permanent hair dyes. The *para*-phenylenediamine is oxidized and the reduced form of the dye reacts with some specific molecule to form the desired color dye.

23.	Aluminum chlorohydrate is an ingredient intended to act as an astringent.

25.	Inhalation can possibly harm nasal passages and lungs.

27.	They work because they stabilize tiny oil particles so they can remain in suspension in the aqueous layer. Without them the oil and water would separate into two layers because oil and water are not soluble in each other.

29.	Cationic detergents are almost always quaternary ammonium halides. An example like hexadecyltrimethylammonium chloride has a positively charged site. Cationic detergents kill bacteria.

Anionic detergents like sodium lauryl sulfate have a negatively charged site.

sodium lauryl sulfate

Nonionic detergents like lauric diethanolamide are electrically neutral.

lauric diethanolamide (an amide detergent)

31. Anionic detergents give good foaming characteristics. The nonionic detergents stabilize foams and thicken the shampoo to give a smoother-pouring liquid. (see question 29)

33. Toothpastes contain abrasives to remove unwanted substances from tooth surfaces and detergents to assist in suspending unwanted particles in water.

35. Optical brighteners fluoresce. They absorb invisible ultraviolet light of high energy and emit light in the visible range.

37. Toilet bowl cleaners are acidic and have a pH less than 2. Drain cleaners are usually bases and have a pH greater than 12. Recall that neutral solutions have pH = 7.

39. An aerosol oven cleaner poses potential threats to the respiratory tract, eyes and skin. Inhalation of droplets of the caustic aerosol can damage nasal passages and lungs by causing chemical burns. The aerosol droplets can reach the eyes; this can permanently harm the lens of the eye because the caustic can denature eye proteins. Skin contact can cause irritation or even chemical burns.

Speaking of Chemistry

Name _____

Consumer Chemistry

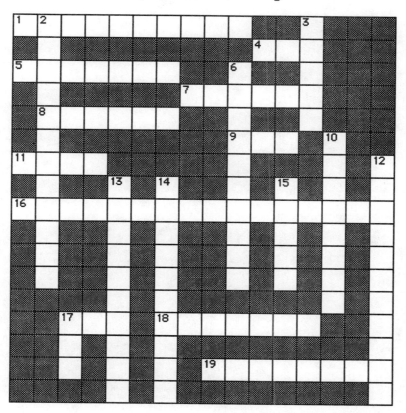

Across

1 French scientist who discovered the bleaching and disinfectant power of chlorine compounds.
4 Food and Drug Administration
5 Pigment produced by the skin when exposed to high energy light.
7 French for froth or lather
8 Calcified plaque that can lead to gum disease.
9 Sun protection factor
11 National Oceanic & Atmospheric Administration
16 Bleaching agent with formula H_2O_2.
17 Ultraviolet light with wavelength greater than 330 nm.
18 Floating soaps are less dense than water because of _____ air.

19 Amino acid that contains -SH group and forms disulfide bridges between protein chains.

Down

2 Procedure for removing hair by using electricity to destroy the hair follicle.
3 High-molecular-weight hydrocarbons as part of formulation in lipstick.
6 Compounds like PABA.
10 Positively charged surfactants which are almost all quaternary amines.
12 Corrosive cleaners like NaOH
13 A mixture of detergents and abrasives.
14 Compounds like sodium sulfide and calcium thioglycolate used to remove hair.
15 Cleaning compounds formed from fatty acids and strong bases.
17 Ultraviolet light with wavelength from 285 nm to 330 nm.

Bridging the Gap   I

Internet access to the Food and Drug Administration, FDA

The FDA home page

The Food and Drug Administration is readily accessible using the Internet. The FDA home page can be reached by using the following universal resource locator (URL) web site address,

http://www.fda.gov/

The square buttons on the FDA home page are interactive. You can click on one of the squares to go to another page that deals with a specific area of FDA activity. The "Cosmetics" button leads to "The FDA Cosmetics Index".

The FDA Cosmetics index

You can get information related to cosmetics from the FDA on the World Wide Web using the following URL. This is the Food and Drug Administration's Cosmetics index. It is maintained by the **Center for Food Safety & Applied Nutrition (CFSAN).**

http://vm.cfsan.fda.gov/list.html

When you access this web site you will see underlined subject headings each leads to another page with a different URL address. The information displayed in the Cosmetics index as of June 6, 1996 is shown on the following page. This page is continuously updated so its appearance will change. Everyone of the underlined lines of text is interactive. You access the next page by clicking on the text. Typically the content under any of these headings can be can be printed or down loaded to disk in the form of text. Your responsibility is to open the page "•Suntan Products, Sunscreens, and Tanning". This page looks similar to following .

> **U.S. Food and Drug Administration**
> **Center for Food Safety and Applied Nutrition**
>
> ### Information about Suntan Products, Sunscreens, and Tanning
>
> **Information from the Center for about Suntan Products, Sunscreens, and Tanning**
>
> - •Suntan Products and Sunscreens
> - •Healthy Tan: A Fast Fading Myth
> - •Drug Photosensitivity Another Reason to be Careful in the Sun

Click on the category "•Healthy Tan: A Fast Fading Myth" it has the following URL.

http://vm.cfsan.fda.gov/~dms/cos-tan.html

Find what the FDA says about the health risks associated with using sun lamps. Try to find what the FDA says about too much sun. Record your findings on the Concept Report Sheet. Open other categories that interest you. Record the URL for the topic you selected, tell why you picked that topic and record your opinion of the FDA material.

Internet access to the
Food and Drug Administration, FDA

FDA Center for Food Safety & Applied Nutrition (CFSAN)
Cosmetics index

COSMETICS

Cosmetic Safety, FDA Requirements and Programs
- Cosmetic Safety and FDA Authority
- Animal Testing
- Color Additives
- Cosmeceuticals
- Inspection of Cosmetics and Imports
- Over the Counter (OTC) Drugs versus Cosmetics
- Prohibited Ingredients
- Reporting Problem Products to FDA
- Shelf Life - Expiration Date
- Voluntary Cosmetic Registration Program

Cosmetic Products that Consumers Frequently Inquire About:
- Animal Grooming Aids
- Aromatherapy
- Artificial Nail Remover
- Eye Products
- Hair Dyes
- Hair Loss Products
- Skin Peeling Products and AHA
- Soap
- Suntan Products, Sunscreens, and Tanning
- Tattooing
- Thigh Creams
- Weight Loss Products

Cosmetic Labeling and Labeling Terms
- Cosmetic Labeling
- Alcohol Free
- Cruelty Free - Not Tested in Animals
- Fragrance Free and Unscented
- Hypoallergenic

Cosmetics of Interest to Specific Groups
- Cosmetics and Teenagers
- Cosmetic Help for Cancer Patients
- Cosmetic Handbook for Industry

Bridging the Gap I
Concept Report Sheet

Name _____

Internet access to the
Food and Drug Administration, FDA

1.   Give the universal resource locator (URL) for the FDA page that discusses
     •Healthy Tan: A Fast Fading Myth

                http://_____

     What does the FDA say about the advisability of sun tanning and exposure to uvA
     and uvB light?

2.   Identify the cosmetic categories you opened.

3.   Give the universal resource locator (URL) for the for the FDA cosmetic category you found
     most interesting.
                http://_____

     Why did you select this category and what is your opinion of the FDA page? Would you
     recommend this URL to someone else? Explain.

Bridging the Gap   II   Name_____
Concept Report Sheet
Personal Hygiene Products

The personal hygiene industry is a multibillion dollar a year business. Products come and go. Styles change and a product that was perfectly satisfactory in 1995 is replaced by another in 1996. What is your opinion of personal hygiene products? Your assignment has the following two parts.

1.   The first is to consult with a group of classmates and settle on reasons why you believe it is beneficial to have many different brands and types of toothpastes, soaps, shampoos, etc. What benefits does this variety offer the consumer? Do you think the variety is of any practical value or is it needless duplication that simply raises the cost of products? What problems does this variety create for the consumer? Do you think the benefits out weigh the problems? Justify your answers.

2.   The second is to describe the possible harm that would result if only one brand of each product were available? For example, why would there be any problem if there were only one brand and formulation of shampoo. Has a lack of product choice ever created a problem for you or someone you know? If it has, explain.

Solution to Speaking of Chemistry crossword puzzle

Consumer Chemistry

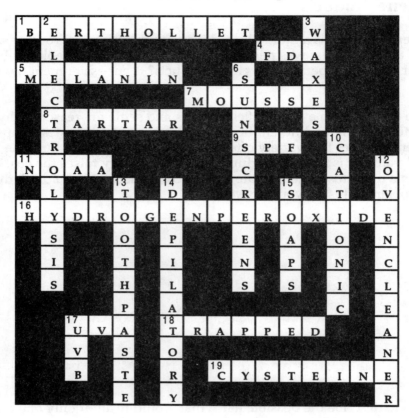

17 Nutrition: The Basis of Healthy Living

17.1 Digestion: It's Just Chemical Decomposition

This section describes the digestion process. The major parts of the digestive system are named and diagrammed. The types of food digested in each part of the digestive system are identified. The products from digestion of carbohydrates, fats and proteins are described.

Objectives

After studying this section a student should be able to:
- name the three major parts of the digestive system
- name the products formed by the digestion of carbohydrates, fats and proteins
- match a type of food with the organ where it is digested
- tell what products result from the digestion of carbohydrates, fats and proteins

Key terms

digestion	mouth	amylase	carbohydrates
starch	stomach	small intestine	protein
lipids	energy production	biomolecules	storage as fat

17.2 Energy: Use It or Store It

This section describes how to calculate a person's BMR.

$$\text{BMR in Calories / day} = 10 \times \text{body weight in pounds}$$

The method for calculating a person's minimum daily caloric needs is shown. The energy available from a gram of fat, a gram of protein and a gram of carbohydrate is described. The activity factors are tabulated for calculating daily caloric need. The process for storing energy in ATP and the release of energy by forming ADP is described.

Objectives

After studying this section a student should be able to:
- give a definition for basal metabolic rate (BMR)
- tell how to estimate a person's BMR using their body weight in pounds
- tell how a person's minimum daily caloric needs are estimated
- tell how many kcal are released by a gram of fat, carbohydrate, or protein
- calculate the minimum daily caloric need of a person from activity level factors and BMR
- calculate the number of kcal that can be obtained from a sample of food if given the grams of food and the percentages for fat, carbohydrate and protein
- tell what ATP and ADP do in the production of energy

Key terms

basal metabolic rate, BMR	Calorie	kilocalorie
fats: 9 kcal/ g	carbohydrates: 4 kcal/ g	proteins: 4 kcal/ g
energy taken in	energy used	energy stored
glucose	ATP	adenosine triphosphate
citric acid cycle	coenzyme	acetyl-coenzyme A

17.3   Sugar and Polysaccharides, Digestible and Indigestible

Carbohydrates are described in terms of digestible and indigestible. Types of digestible carbohydrates are listed. Specific indigestible carbohydrates are identified. Dietary fiber is defined and its importance is explained. Carbohydrate content for some common foods is tabulated in Table 17.4.

Objectives
After studying this section a student should be able to:

   name three kinds of digestible carbohydrates
   identify two indigestible carbohydrates
   name sources of insoluble fiber and soluble fiber
   describe how insoluble fiber influences risks of heart disease and colorectal cancer

Key terms

simple sugars	glucose	fructose
disaccharides	sucrose	maltose
polysaccharides	amylose	amylpectin
indigestible carbohydrates	cellulose	pectin
dietary fiber	insoluble fiber	soluble fiber
heart disease	colorectal cancer	

17.4   Lipids   Mostly Fats and Oils

This section describes the sources of dietary fats. The structure and breakdown of stored body fat is discussed. Heart disease (atherosclerosis) is described. The link between heart disease and fats is explained. The formation of lipoproteins from cholesterol is described. The role of HDL and LDL in heart disease is explained. Alternate ways to reduce dietary fat intake with fat substitutes are discussed.

Objectives
After studying this section a student should be able to:

   name the common sources of dietary fats
   tell what two types of compounds result from hydrolyzing fats
   tell why high fat intake increases the risk of heart disease
   give a description of heart disease
   describe the connection between cholesterol and lipoproteins
   give definitions for HDL and LDL
   tell how good and bad cholesterol differ relative to the risk of heart disease

Key terms

lipids	fats	oils
triglycerides	glycerol	free fatty acids
saturated fat	monounsaturated fat	polyunsaturated fat
atherosclerosis	plaque	heart attack
cholesterol	lipoproteins	LDLs
HDLs	"bad" cholesterol	"good" cholesterol
fat substitutes	"Simplesse™ "	

17.5   Proteins in the Diet

This section identifies high protein foods. The function of dietary protein is explained. The fate of excess dietary protein is described.

Objectives
After studying this section a student should be able to:
> identify the major dietary sources of protein
> describe the role of dietary protein
> tell what the body does with excess protein
> identify and describe a protein deficiency disease

Key terms
high-protein foods	amino acids	excess amino acids
urea	ammonia	

17.6   Our Daily Diet
This section describes the Food and Drug Administration's (FDA) food triangle that is displayed on some food labels. Micronutrients and macronutrients are defined. The information listed on a Nutrition Facts Label is explained. The mandatory listings in a Nutrition Facts Label are identified. Health risks are described for dietary deficiencies of iron, calcium, iodine, and other nutrients.

Objectives
After studying this section a student should be able to:
> explain the purpose of the FDA food pyramid
> give a definition for an empty calorie
> name the nutrients that are in the mandatory listings on the Nutrition Facts Label
> describe the basis of the "daily reference values" for nutrients
> explain why iron, calcium, vitamin A, and vitamin C are in the mandatory list
> give definitions for osteoporosis, anemia

Key terms
food pyramid	"empty calories"	Nutrition Facts label
% Daily Value	macronutrients	micronutrients
daily reference values, DVs	mandatory listings	anemia
osteoporosis	cancer risk	

17.7   Vitamins in the Diet
This section gives a definition for vitamins. The water-soluble and fat-soluble classes of vitamin are described. The structures and functions of specific vitamins are summarized. Typical sources of vitamins A, D, E, K, B, and C are tabulated.

Objectives
After studying this section a student should be able to:
> give the definition for vitamins
> explain why some vitamins are water-soluble and others are fat soluble
> tell why the solubility classification of vitamins is important
> tell which types of vitamins are stored in the body and which are not
> tell what is meant by a vitamin megadose
> name some typical water-soluble vitamins and some typical fat-soluble vitamins
> describe the functions of vitamins such as A, D, E, K, B and C
> explain what an antioxidant does and name a vitamin that is an antioxidant
> give a definition for a coenzyme
> name a health problem that results from a deficiency of each vitamin A, D, E, K, B and C

Key terms

vitamins	megadoses	fat-soluble
nonpolar	vitamins A, D, E, and K	antioxidant
night blindness	rickets	water-soluble
coenzymes	B vitamins	pellagra
vitamin C, ascorbic acid	scurvy	placebo

17.8 Minerals in the Diet

This section defines and identifies mineral nutrients. Macrominerals and microminerals are identified. The role of minerals in electrolyte balance is explained. Sources of each mineral are tabulated. Mineral deficiency diseases are identified: calcium and osteoporosis; iodine and thyroid function; iron and anemia.

Objectives

After studying this section a student should be able to:

name sources for Na^+, K^+, Cl^-, Ca^{2+}, P, Mg^{2+} and S

describe the role of each mineral: Na^+, K^+, Cl^-, Ca^{2+}, P, Mg^{2+} and S

tell what mineral is needed to prevent osteoporosis

describe the role sodium ion concentration in high blood pressure

describe the health problems that result from deficiencies in iodine, iron, calcium and zinc

tell what mineral is typically deficient when a person has anemia

Key terms

minerals	macronutrient minerals	calcium
phosphorous	magnesium	sodium
potassium	chlorine	sulfur
electrolyte balance	normal blood pressure	high blood pressure
osteoporosis	micronutrient minerals	iron
copper	zinc	iodine
thyroid	anemia	

17.9 Food Additives

This section defines food additives. The function and range of food additives is described. The Food and Drug Administration (FDA) GRAS list is explained. The specific functions of food additives are defined: preservatives, antioxidants, sequestrants, flavorings, flavor enhancers, food colorings, pH adjusters, anticaking agents, stabilizers and thickeners. Examples of each type of food additive are identified. The Delaney Clause in the 1958 FDA Act is discussed. The legal principle of *de minimis* is described.

Objectives

After studying this section a student should be able to:

explain why food additives are necessary

tell what is meant by the GRAS list

explain why foods can be preserved by drying

describe the way that concentrated salt and sugar solutions preserve foods

name two antioxidants and tell why they are added to foods

tell why sequestrants are added to food

explain why food flavorings and flavoring enhancers are added to food

tell why BHA CH_3-O and BHT are added to food
tell why pH adjusters are food additives
describe the Delaney Clause in the 1958 FDA Act
explain what is meant by the *de minimis* principle

Key terms

food additives	GRAS list	preservatives
drying	dehydrate	osmosis
sodium benzoate	sodium propionate	salted
antioxidants	rancid	BHT
BHA	sequestrants	complexes
food flavors	volatile oils	extracts
flavor enhancers	potentiators	MSG
food colors	pH control	buffers
preservatives	antioxidants	viscosity modifiers
anticaking agents	hygroscopic	magnesium silicate
stabilizers	thickeners	hydroxyl groups
Delaney Clause	*de minimis*	

17.10   Some Daily Diet Arithmetic

This section shows how to calculate the percent fat in a food using the information on a Nutrition Facts Label. An example calculation shows how to calculate the caloric value of a food from the grams of fat, grams of carbohydrate and grams of protein. An example calculation is given that shows how to determine the number of kcal and grams of fat, protein, and carbohydrate a person should consume to meet their daily calorie requirement and stay in the recommended percentages of fat, carbohydrates and protein.

Objectives

After studying this section a student should be able to:
read a Nutrition Facts Label to find the grams of fat, carbohydrate and protein in the food
read the label to find the total Calories and the Calories from fat and then calculate
       % Calories from fat
read a food label to find the grams of fat, protein and carbohydrate, calculate the total
       Calories, and the % Calories from fat
calculate the %Calories from fat if given total Calories and number of Calories from fat
calculate the Calories from fat, protein, and carbohydrates from a total Calories a
       recommended 30% from fat, 10% from protein and 60% from carbohydrates
state the number of dietetic Calories available from 1 gram of fat
state the number of dietetic Calories available from 1 gram of protein
state the number of dietetic Calories available from 1 gram of carbohydrate

Key terms

Nutrition Facts labels total calories from fat

$$\% \text{ Calories from fat} = 100 \times \left(\frac{\text{Calories from fat}}{\text{Total calories}} \right)$$

$$\frac{9 \text{ kcal}}{\text{g fat}} \qquad \frac{4 \text{ kcal}}{\text{g protein}} \qquad \frac{4 \text{ kcal}}{\text{g carbohydrate}}$$

Additional readings

Benson, Kim D. "Fast Fat" ChemMatters Feb. 1990: 13.

Heaney, Robert P. " Protein Intake and the Calcium Economy" Journal of the American Dietetic Association 22 Nov. 1993, 1259.

Hunter, Beatrice T. " How Safe are Nutritional Supplements?" Consumers' Research Magazine Mar. 1994: 21.

Hunter, Beatrice T. " Small but Significant (role of trace elements in human health)" Consumers' Research Magazine Dec. 1994: 8.

Sangiorgio, Maureen et al. "A Salty Surprise (sodium sources)" Prevention Feb. 1992: 12.

" Taking Vitamins: Can They Prevent Disease?" Consumer Reports Sept. 1994: 561.

Tarnopolsky, Mark. "Protein, Caffeine, and Sports: Guidelines for Active People" The Physician and Sportsmedicine March 1993: 137.

"The Facts About Fats" Consumer Reports June 1995: 389.

Wurtman, Richard J. "Nutrients That Modify Brain Function" Scientific American April 1982: 50.

Yulsman, Tom. "HDL Finally gets a Hearing" Medical World News March 1992: v33 n3 27.

Brown, M. H. "Here's the Beef: Fast Foods Are Hazardous to Your Health" Science Digest, 1986: 94, No. 4, 30.

Pond, C. "Fat and Figures" New Scientist, June 4, 1987: 62-66.

Answers to Odd Numbered Questions for Review and Thought

1. Digestion is the process of breaking large molecules in food into substances small enough to be absorbed by the body from the digestive tract.

3. a. Fats: 9 kcal/ g or 9 Calories/ g
   b. Proteins: 4 kcal/g or 4 Calories/g
   c. Carbohydrate: 4 kcal/ g or 4 Calories/g

5. Metabolic energy is the energy expended to keep the heart beating, the lungs inhaling and exhaling air, the nerves generating and transmitting their flood of electrical impulses, and the cells of the body conducting their normal functions. The basal metabolic rate, BMR, is the rate at which the body uses energy to support these maintenance operations. It is typically in kcal or Calories per hour. It is the minimum energy required for a person to stay alive.

7. A triglyceride is a triester formed from glycerol and three fatty acids.

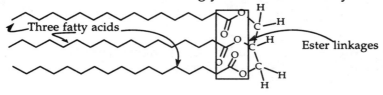

9. a. Atherosclerosis is a disease that is associated with the buildup of fatty deposits on the inner walls of arteries, i.e. hardening of the arteries, notably in the brain and in the heart.. These deposits lead to higher risks of heart attacks.
   b. Cholesterol is a lipid steroidal alcohol that contributes to the development of artherosclerosis.
   c. Plaque is the yellowish deposit of cholesterol and lipid-containing material on artery walls that is a symptom of artherosclerosis. These deposits can promote blood clots that cut off blood supply and cause heart attacks and strokes.
   d. Lipoproteins are complex assemblages of lipids, cholesterol and proteins that serve to transport water-soluble lipids in the blood stream through the body.
   e. Low density lipoproteins are richer in lipid than in protein and have a corresponding lower density in the range of 1.006-1.063 g/mL.
   f. High density lipoproteins are richer in protein than in lipid with a corresponding higher density in the range of, 1.063-1.210 g/mL.

11. Empty calories provide energy but very few, if any, vitamins, minerals or proteins. A nutritious calorie is derived from a carbohydrate that is part of a food that provides the same energy and is accompanied by protein, vitamins or minerals. The most notorious empty-calorie foods are sugar, fat, and alcohol.

13. The major categories of information required on a Nutrition FactsLlabel are the total Calories, Calories from fat, the weight in grams or milligrams per serving plus the % Daily Values for macronutrients, total fat, saturated fat, unsaturated fat, cholesterol, sodium, total carbohydrate, dietary fiber, sugars and protein.

15. Micronutrients are minerals and vitamins, needed in small amounts. Four minerals are allowed to be listed: iron (Fe), copper (Cu), zinc (Zn) and iodine (I). It is mandatory to list only the micronutrient mineral, iron (Fe).

17. Overdoses of fat-soluble vitamins are usually more of a problem than water-soluble vitamins. Fat-soluble vitamins can be stored in the fatty tissues of the body. The liver is one of the main sites. Fat-soluble vitamins have nonpolar hydrocarbon structures, so they are similar to the fat molecules. Megadoses can cause varying adverse physiological problems. Water-soluble vitamins are not stored in the body, excesses are excreted in the urine. They contain polar groups so they are attracted to the polar water molecules.

a. Riboflavin is a water-soluble vitamin; excess amounts are excreted

Riboflavin, polar groups

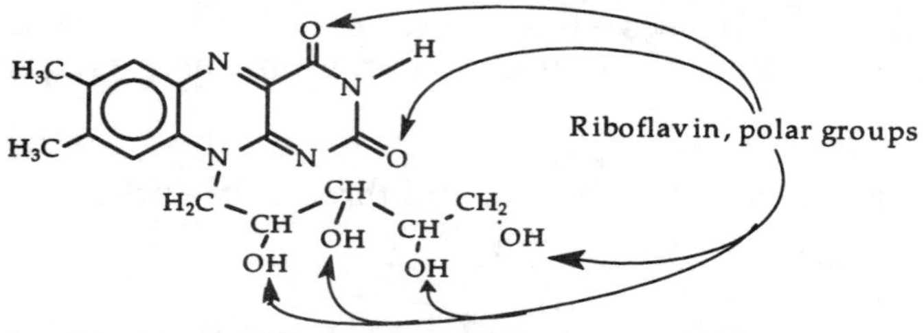

b. Vitamin C is a water-soluble vitamin; excess amounts are excreted. Megadoses can cause diarrhea, nausea, and abdominal cramps.

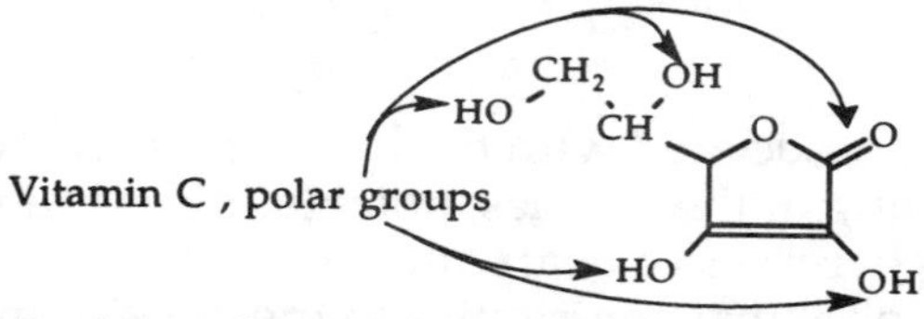

Vitamin C, polar groups

c. Vitamin A is a fat-soluble vitamin; excess amounts are stored in fatty tissue. An overdose can cause headache, fatigue and vomiting.

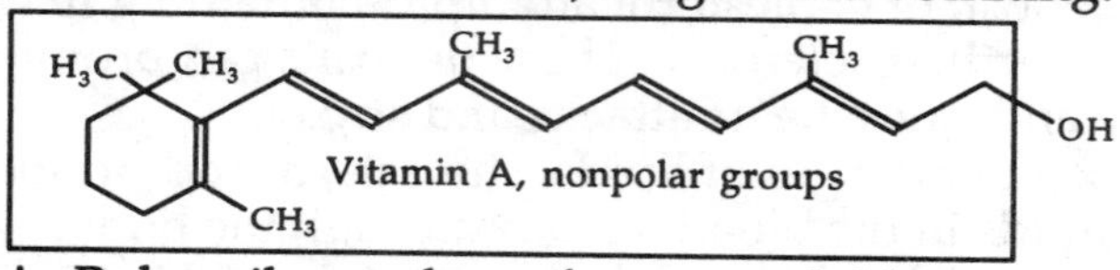

Vitamin A, nonpolar groups

d. Vitamin B describes a class of water-soluble vitamins; excess amounts are excreted, but megadoses can cause nerve damage and even paralysis.

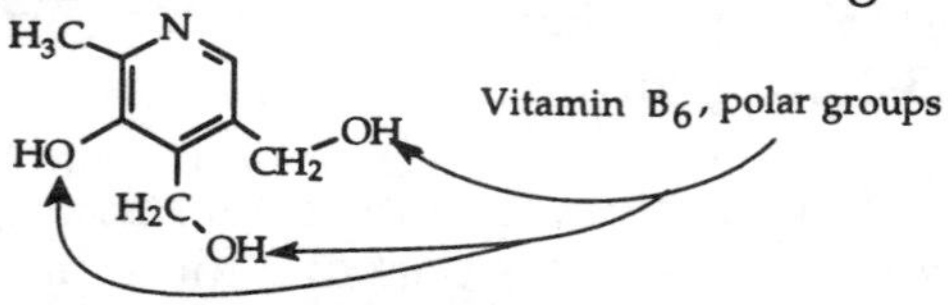

Vitamin B_6, polar groups

e. Vitamin D is a fat-soluble vitamin; excess amounts are stored in fatty tissue. Overdosage can lead to calcium deposits in the kidney, lungs and tympanic membrane of the ear. Deafness can result.

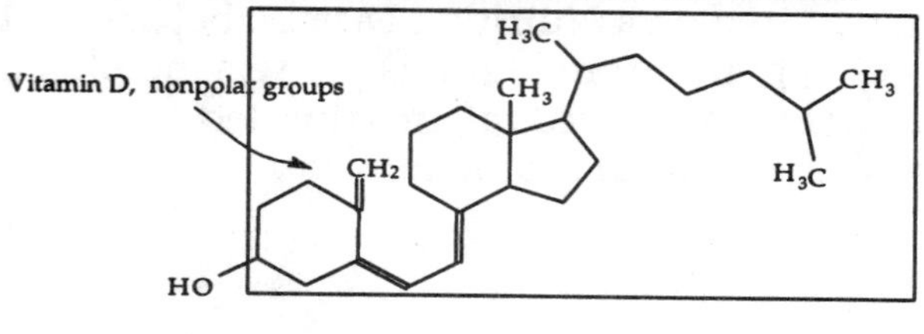

Vitamin D, nonpolar groups

19. An electrolyte is a substance that dissolves in water to produce ions. The electrolyte balance refers to a condition of proper transfer of material through osmosis, normal nerve impulse transmission, normal acid-base balance and extracellular volume.

21. The seven most abundant macronutrient minerals in the body are phosphorus, calcium, magnesium, sodium, potassium, chlorine, and sulfur.

23. a. Sodium benzoate, preservative
    b. Sodium propionate, preservative
    c. BHA (butylated hydroxyanisole), antioxidant
    d. BHT (butylated hydroxytoluene), antioxidant
    e. Citric acid, sequestrant
    f. Sodium EDTA, sequestrant

25. a. Monosodium glutamate, flavor enhancer   b. Yellow No. 5, food color
    c. Acetic acid, buffer or pH control   d. Calcium silicate, anticaking agent
    e. Carrageenan, thickener   f. Lecithin, antioxidant

27. Lactose is a disaccharide because it can be broken down into two monosaccharides. When hydrolyzed lactose can be broken down into two monomers. These are D-galactose monomer and D-glucose monomer.

29. The volume of these globules would be related by the ratio of the energy equvalents of 9 kcal per gram of fat and 4 kcal per gram of carbohydrate. The body would still need to store the same amount of energy but the mass of storage material would increase by a factor of 9/4 or 2.25 times. The space required to store the same amount of energy in would more than double.

31. The question each person considers important will differ. Here are some possible questions. Are there other ways to make adequate folic acid doses available? What are the effects of a folic acid overdose? What are the costs of treating overdose victims?

33. The program might benefit individuals who would suffer from osteoporosis. A negative side to the plan is that some individuals who are prone to kidney stones would face greater chances of forming kidney stones if their dietary calcium intake increased. Another possible negative would be that the plan would increase the reach of government into the citizen's daily life. This medication of the entire population might undermine individual responsibility for personal health.

35. Fats and fatty acids undergo oxidation when they become rancid. Oxidation is enhanced by the presence of trace metals. Sequestrants bind the metal ions and decrease the extent of oxidation.

37. There is no right answer for this question. Each answer will depend on the specific product.

39. A "zero-risk" law allows for no risk of harm and does not weigh the risks compared to the benefits. It is a direct ban of a substance that poses an identified risk.

41. The answer to this question clearly depends on the specific group.

Answers for Selected Problems

1. In this case BMR = 10 x weight of 110 pounds and the activity factor is 1.6.
   10 x 110 x 1.6 =1760 Calories needed daily

3. Bicycling uses 5 Calories for each minute of riding time, or in 1 minute a person will consume 5 Calories. Using 100 Calories will require 20 minutes of riding time.

$$100 \text{ Calories} \times \frac{1 \text{ min}}{5 \text{ Cal}} = 20 \text{ minutes}$$

7. Since we know that the dangerous cholesterol level is 240 mg per 100 ml of blood, we need to express the person's blood volume in milliliters; we need to change 13 pints to milliliters. Then we can multiiply by the number of mg of cholesterol in 100 ml of blood to get the person's total amount.

$$13 \text{ pt} \times \frac{473 \text{ ml}}{1 \text{ pt}} \times \frac{240 \text{ mg cholesterol}}{100 \text{ ml blood}} = 14758 \text{ mg} = 15000 \text{ mg}$$

   This is 15 grams. Round to 2 significant digits because 13 only has 2 sd.

11. Total Calories are found by adding together the Calories from the protein, carbohydrate, and fat; protein and carbohydrate have 4 Cal. per 1 gram while fat has 9 Cal. per gram.

$$\text{Calories from protein} = 34 \text{ g} \times \frac{4 \text{ Cal}}{1 \text{ gram}} = 136 \text{ Calories}$$

$$\text{Calories from carbohydrate} = 44 \text{ g} \times \frac{4 \text{ Cal}}{1 \text{ gram}} = 176 \text{ Calories}$$

$$\text{Calories from fat} = 46 \text{ g} \times \frac{4 \text{ Cal}}{1 \text{ gram}} = 414 \text{ Calories}$$

   136 + 176 + 414 = 726 Calories total

12. First calculate her daily Calorie need; since she is extremely active, her activity factor is 1.9.
   BMR = 1450 x 1.9 = 2755 Calories for her daily Calorie need.
   Now find 30% of 2755 to give the number of Calories to come from fat, 60% of 2755 to get the Calorie from carbohydrates and 10% of 2755 to get the Calories needed from protein. Finally convert each number of Calories to grams of substance using 1 gram per 4 Cal for carbohydrate and protein and 1 gram per 9 Cal for fat..

$$\text{grams of fat} = .30 \times 2755 \text{ Cal} \times \frac{1 \text{ gram}}{9 \text{ Cal}} = 92 \text{ g}; \text{grams of carbohydrate} = .60 \times 2755 \text{ Cal} \times \frac{1 \text{ gram}}{4 \text{ Cal}} =$$

$$413 \text{ g}; \text{grams of protein} = .1 \times 2755 \text{ Cal} \times \frac{1 \text{ gram}}{4 \text{ Cal}} = 69 \text{ g}.$$

Speaking of Chemistry

Name _____

The Basis of Healthy Living

Across

3 _____ prevents goiter.
4 Pellagra results from a deficiency of
7 The _____ acid cycle produces energy from molecules that contain two carbon atoms.
9 Vegetable fats.
12 The buildup of fatty deposits on artery walls.
13 Adequate _____ prevents osteoporosis.
14 Adenosine triphosphate or _____
16 The generally recognized as safe list.
19 A chelating agent used as a sequestrant.
20 The _____ clause requires use of substances that pose zero cancer risk.

Down

1 Fats and oils are both _____
2 Monosaccharides are simple _____.
5 The waxy lipid linked to heart disease.
6 Indigestible polysaccharides.
8 The material deposited on walls of arteries that results in heart disease.
10 The primary monosaccharide energy source.
11 Ascorbic acid is known as _____ C.
15 A polypeptide containing many amino acids.
17 Produce 9 C / g when metabolized.
18 Low density lipids.

270

Bridging the Gap
Concept Report Sheet

Name _____

The Delaney Clause and the "*de minimis*" principle

The quality and safety of food and medicine has been a concern for people since the beginning of civilization In Western Europe, in 1202 AD, King John of England proclaimed the first English food quality law, the Assize of Bread, which prohibited the adulteration of bread with ingredients like ground peas or beans. Regulations in the United States date back to the colonial period. The State of Massachusetts enacted the first general food adulteration law in the United States in 1848. The need for regulation is accepted. The issue is, "How much regulation?".

This exercise has three parts. You may do this alone or as part of a discussion group. One part is to write a brief argument for continuing to follow the Delaney Clause. Another part of your assignment is to write an argument for following the "de minimis" principle. Lastly you are to write two questions you would want answered before you would make a final decision about this issue.

1.   Argument for following the Delaney Clause

2,   Argument for following the *de minimis* principle

3.   Your questions regarding the issues.

Solution to Speaking of Chemistry crossword puzzle
The Basis of Healthy Living

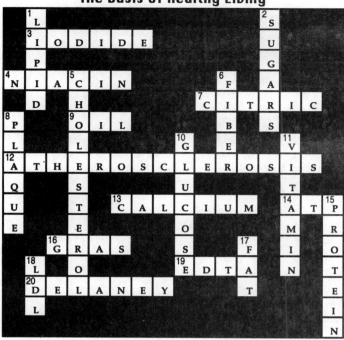

18   Toxic Substances

18.1   The Dose Makes the Poison

This section defines the terms: toxic substance, poison, dose, lethal dose, and LD_{50}. Lethal doses for some common substances like aspirin and ethanol are listed. Difficulties in establishing LD_{50} values are discussed. Species differences in LD_{50} for the same compound are explained. LD_{50} values are tabulated for a variety of compounds from aspirin to lead and nicotine. Three routes of entry to the body are identified: inhalation, ingestion, and skin contact.

Objectives

After studying this section a student should be able to:

give definitions for poison, chronic toxicity and acute toxicity

give definitions for dose, lethal dose, sublethal dose, LD_{50}

explain how human lethal doses can be established without human testing

explain why it is difficult to estimate risk to humans using LD_{50} values from animal tests

calculate the total dose that would pose a 50% chance of killing a subject given body mass and LD_{50}

tell which compound in a set poses the greatest risk if given the LD_{50} values for each

name the six classifications of toxic substances

Key terms

dose	poison	lethal dose
LD_{50}	species differences	animal data
milligram / kilogram	mg/kg	modes of entry
inhalation	ingestion	skin contact
acute toxicity	chronic toxicity	corrosive poison
metabolic poison	neurotoxic poison	mutagenic poison
teratogenic poison	carcinogenic poison	

18.2   Corrosive Poisons

This section describes corrosive poisons which typically destroy tissue. Specific corrosive poisons are identified. The reaction of strong acids is illustrated for attack on a protein. The warning properties of corrosive poisons are described. Emergency procedures to deal with corrosive poison exposures are described.

Objectives

After studying this section a student should be able to:

give the definition for corrosive poisons

give examples of corrosive poisons

explain what is meant by a "warning property"

give the definitions for strong acids and strong bases

tell what is meant by a mineral acid

describe how a strong acid reacts with peptide bonds to damage protein

describe the warning properties of ammonia

explain why hydrofluoric acid is especially hazardous

name a corrosive poison that reacts in the lungs to produce pulmonary edema tell how the corrosive poisons ozone, hydrogen peroxide, and the halogens attack tissue

describe the emergency steps to take when exposed a corrosive poison

Key terms

corrosive poisons	strong acids	strong bases
hydrochloric acid, HCl(aq)	sulfuric acid, H_2SO_4(aq)	sodium hydroxide, NaOH
mineral acids	acid-catalyzed hydrolysis	dehydration
warning properties	base-catalyzed hydrolysis	pulmonary edema
oxidizing agent	hydrogen peroxide	ozone
fluorine	chlorine	halogens
reducing agent	sulfite ion	oxalic acid

18.3 Carbon Monoxide and Cyanide as Metabolic Poisons

Metabolic poisons are defined. The specific metabolic poisons carbon monoxide, :C:::O:, and cyanide ion, :C:::N:⁻, are discussed in detail. Acute toxic effects and subacute chronic effects of metabolic poisons are described. The reactions that make CO and CN⁻ toxic are illustrated and explained. Sources of CO poisoning like automobile exhaust and kerosene heaters are identified. Symptoms of CO poisoning are described. Treatment for CO poisoning is discussed. The warning property of CN⁻, the faint odor of almonds, is described. The natural sources of CN⁻ such as cherry seeds are summarized. The slow natural mechanism for purging CN⁻ from the body is described. Antidotes like thiosulfate, $S_2O_4^{2-}$, are identified.

Objectives

After studying this section a student should be able to:

give the definition for metabolic poisons

name two examples of metabolic poisons

summarize the way that CO interferes with oxygen transport

explain why CO poisoning is reversible and how it can be treated

describe permanent effects of CO poisoning

name the sources of CO

name processes that produce cyanide and natural sources of cyanide ion, CN⁻

describe the warning properties of CO and CN⁻

identify the molecule in the body that reacts with CO

identify the molecule in the body that reacts with CN⁻

describe what treatment or antidotes are used in case of CO and CN⁻ poisoning

Key terms

metabolic poisons	chronic effects	carbon monoxide, CO
oxygen transport	hemoglobin	oxyhemoglobin
carboxyhemoglobin	bond strength	combustion
incomplete combustion	complete combustion	reversible reaction
Le Chatelier's principle	equilibrium shift	cumulative poison
cyanide ion, CN⁻	odor of almonds	natural occurrence
acid hydrolysis of amygladin	asphyxiation	cytochrome oxidase
antidotes	rhodanse	thiosulfate

18.4 Heavy Metals as Metabolic Poisons

This section discusses heavy metal metabolic poisons that react with sulfhydryl (-SH) groups in enzymes. Common heavy metal poisons like arsenic (As), lead (Pb), and mercury (Hg) are described. Chelate antidotes for heavy metal poisoning are illustrated. The action of heavy

276

metals on sulfhydryl groups is illustrated. Definitions and examples are given for amalgams, chelating agents, British anti-Lewisite, lead poisoning, and pica. The federal Center for Disease Control, CDC, is identified. An example of the calculation of dose using ppm and ppb concentrations is presented. Reactions between chelating agent antidotes and heavy metals are explained. Public health agency efforts to prevent heavy metal poisoning are described.

Objectives
After studying this section a student should be able to:

give a definition for heavy metal poisons
name three common heavy metal poisons
explain why heavy metals are toxic
give a definition for chelating agents and tell how they relate to heavy metal antidotes
draw the structure for a sulfhydryl group
name sources for arsenic, lead, and mercury
describe the symptoms of lead poisoning
tell how heavy metal poisoning can be treated

$$\text{H-OCH}_2\text{-CH(SH)-CH}_2\text{-S-H}$$

give the name for the compound with the structure
tell what the Center for Disease Control (CDC) intervention level for lead means
give the definition for amalgam
tell which forms of arsenic and mercury are most toxic

Key terms

heavy metal poisons	cadmium	chromium
nickel	copper	arsenic
high atomic weights	sulfhydryl groups, -SH	Lewisite
BAL	chelating agent	sequestrants
mercury	mercury (I) ion	mercury (II) ion
amalgam	lead	$\mu g/L$
bone marrow	pica	paints
blood levels	$\mu g/dL$	deciliter
Centers for Disease Control, CDC	intervention level	parts per billion, ppb
chelating agent	$Na_2[Ca(EDTA)]$	EDTA

18.5   Neurotoxins

Neurotoxins are defined and common examples like curare and strychnine are identified. The process of nerve impulse transmission is discussed. The roles of acetylcholine and acetyl-cholinesterase in impulse transmission are described. Anticholinesterase poisons are discussed. The symptoms of anticholinesterase poisoning are described. Specific anticholinesterase poisons are illustrated. Alkaloid neurotoxins (alkaloids) like curare are discussed. The physiological effects of alkaloids are described.

Objectives
After studying this section a student should be able to:

give a definition for neurotoxins, neurotransmitter, synapse
describe how a neurotoxin interferes with normal nerve transmission
name two types of neurotoxins
describe how anticholinesterase poisons interfere with nerve transmission
describe how curare and atropine act as neurotoxins

describe the symptoms of anticholinesterase poisons
name two anticholinesterase poisons
describe the symptoms of alkaloid neurotoxin poisoning
name beneficial uses for alkaloids
identify beneficial uses for organophosphates

Key terms

neurotoxins	strychnine	synapse
neurotransmitter	acetylcholine	choline
anticholinesterase poisons	organic phosphates	cholinesterase
insecticide, Sevin	alkaloids	atropine
curare	nerve gases	tabun
sarin		

18.6   Teratogens

Teratogens are defined. The vulnerability of an embryo to teratogens during the different periods of pregnancy and the risks to a fetus during pregnancy are explained. The infamous thalidomide birth defects tragedy is recounted. The importance of molecular structure in teratogenic activity is explained. Teratogenic substances and their effects are tabulated in Table 18.8.

Thalidomide

Objectives

After studying this section a student should be able to:
give the definition for teratogen
give the approximate percentage of birth defects caused by known teratogens
identify the three periods during pregnancy when teratogens pose the greatest threat to a fetus
name the period during pregnancy when a fetus is most vulnerable to a teratogen
describe the circumstances and results of the thalidomide birth defects case
name common teratogenic substances

Key terms

teratogens	chemicals	high-energy radiation
viral agents	organogenesis	thalidomide
chiral	enantiomer	fetal period

18.7   Mutagens

Mutagens are defined and their general properties are described. Difficulties in identifying mutagenic substances in humans are discussed; the impact of the time period for a generation is also considered. The reaction of nitrous acid with NH_2 groups in DNA bases is described and illustrated. The Ames test for mutagen screening is explained. Proven mutagens like ozone, caffeine and acetone are listed in Table 18.9

Objectives

After studying this section a student should be able to:
give the definition for mutagen
tell why it is difficult to gather data on mutagenic effects in humans
describe the results of nitrous acid on bacteria

explain why the FDA is discouraging the use of sodium nitrate as a food preservative
describe the Ames test and tell what it tests
name some common sources of mutagens

Key terms

mutagens	DNA altering	germinal cells
mutation	chromosome alteration	nitrous acid, HNO_2
LSD	ozone	caffeine
benzo(α)pyrene	adenine	guanine
cytosine	amino group	sodium nitrite, $NaNO_2$
food preservative	somatic cells	botulism
Ames test	histidine	

18.8 Chemical Carcinogens

Chemical carcinogens are defined. The three ways cancer manifests itself are described:
1. abnormal rate of cell growth
2. spread of cells to other tissue
3. partial or complete loss of specialized functions

Carcinogenesis is described as a two step process of initiation and promotion. The metastasis or spread of cancer is discussed. The testing procedures to screen chemicals for carcinogenicity are described. Potential flaws are pointed out in the methodology using a maximum tolerated dose (MTD) applied to genetically uniform laboratory animals. Common natural carcinogens in foods like coffee and bananas are identified and listed in Table 18.11.

H_3C, H_2C=C—〈 〉—CH_3 D-Limonene in citrus fruits

Objectives

After studying this section a student should be able to:
give the definition for a carcinogen
name the two stages in carcinogenesis
describe the initiation stage in carcinogenesis
describe the promotion stage in carcinogenesis
give the definition for metastasis
give the definitions for carcinomas, lymphomas, leukemias, and sarcomas
explain why the choice of dose is a problem in testing compounds for carcinogenicity
tell why increasing life expectancies can lead to increased cancer deaths
describe the shortcomings in current carcinogenicity testing
name three common foods that contain natural carcinogens
explain why the use of the maximum tolerated dose (MTD) is a potential flaw in
       carcinogenicity testing
explain why carcinogenicity testing using specially bred laboratory animals may be a
       poor practice
why it is probably impossible to completely avoid contact with carcinogens
tell what the purpose Delaney Clause in the Food and Drug Act of 1958 serves

Key terms

carcinogens	rate of cell growth	spread to other tissues
loss of specialized function	aromatic hydrocarbon	coal dust
initiation	promotion	carcinomas
lymphomas	leukemias	metastasis
animal tests	extrapolation	genetically uniform

sarcoma
chronic mitogenesis
black pepper, saffrole
threshold dose of toxicity

Delaney Clause
chronic wounding
hydrazines

maximum tolerated dose, MTD
natural carcinogens
mustard

Additional readings

Adams, Elijah. "Poisons" Scientific American Nov. 1959: 76.

Hunter, B. T. "Iron Poisoning in Young Children" Consumer's Research Magazine April 1995: 8.

Newman, Alan. "Ranking Pesticides by Environmental Impact" Environmental Science & Technology July 1995: 324A.

Laliberte, Richard. "Is Your Home Hazardous?"(includes related article on carbon monoxide poisoning) Parents Magazine July 1994: 34.

Snyder, James D. "Off-the-shelf bugs hungrily gobble our nastiest pollutants" Smithsonian Magazine April 1993: 66.

Selavka, C. M. "Poppy-seed Eaters May Flunk Morphine Test" Chemical & Engineering News **1994**, 72(32), 56.

Weaver, Daniel C. "Heavy Metal. (a case of mercury poisoning)" Discover April 1993: 76.

Fackelmann, Kathy A. "Can Dental Fillings Create Drug Resistance?" Science News April 10, 1993: 230.

"PHS Reports on Dental Amalgam (Public Health Service)" FDA Consumer April 1993: 389.

Rauber, Paul. "Mercury Madness"(mercury in fish) Sierra Nov-Dec 1992: 36.

Weinberg, Robert A. "A Molecular Basis Of Cancer" Scientific American Nov. 1983: 126.

Wolfe, Morris. "Dental Flaws (the controversy surrounding mercury amalgam fillings)" Saturday Night Nov. 1993: 15.

Answers to odd numbered Questions for Review and Thought

1.
   a. A dose of a toxic substance is the amount of substance that enters the body of the exposed organism
   b. The LD_{50} is the dose that is lethal to 50% of the population tested.
   c. Poisons are substances that are dangerous (possibly toxic) in small amounts.
   d. Warning properties are interactions of a substance with the senses to let a person know of exposure to a poison.
   e. Pulmonary edema results from the collection of fluid in the lungs because an inhaled poison draws water from the surrounding tissue.

3. a. Any molecule or ion that can bind to the same metal ion at two or more sites to form a water-soluble complex ion is a chelating agent.
   b. An amalgam is an alloy of mercury and another metal. Dental amalgams are silver-mercury alloys.
   c. Pica is a craving for non-food items including flaking paint, clay, ice. This frequently causes lead poisoning when the flaking paint contains lead.
   d. A metabolic poison is one that can cause illness or death by interfering with metabolic chemical reactions. These usually react with the sulfhydryl (-SH) groups in enzymes.
   e. An antidote is a molecule that can counteract the effects of a poison by destroying it or rendering it ineffective.

5. a. A mutagen is a chemical that can change the hereditary pattern of a cell, one that is capable of altering the structure of DNA.
   b. A carcinogen is a substance that causes cancer.
   c. Cancer is a condition that exists when cells have abnormal or out-of-control cell growth, are invasive, and lead to a loss of specialized function of cells in the invaded tissue.
   d. The MTD is the maximum tolerated dose, which is almost but not quite a toxic dose.

7. Toxic substances can enter the body through skin contact, ingestion, and inhalation. Inhalation allows the quickest contact with the blood stream in the lungs.

9. Hydrochloric acid causes pulmonary edema while CO and CN^- do not. Mineral acids like HCl have a strong attraction for water. If HCl is inhaled it enters the lungs and draws water from surrounding tissue into the lungs. The water dilutes the acid and collects producing pulmonary edema. The CO and CN^- are not attracted to water as much as the HCl is. The CO is strongly attracted to the Fe in hemoglobin and the CN^- is strongly attracted to the Fe in enzymes like cytochrome oxidase so neither is expected tie up water and produce edema.

11. A corrosive poison would be expected to cause eye and skin damage.

13. Carbon monoxide and cyanide are both metabolic poisons. Carbon monoxide binds to hemoglobin and keeps it from picking up oxygen because CO is more strongly bound than O_2. Cyanide binds to iron in oxidative enzymes like cytochrome oxidase, so the use of oxygen for oxidation in cells is disrupted.

15. Carbon monoxide binds to the iron in hemoglobin.

17. Cyanide binds to iron in oxidative enzymes like cytochrome oxidase.

19. Carboxyhemoglobin has carbon monoxide bound to the iron in hemoglobin. Normal hemoglobin contains iron attached to O_2 or CO_2. Neither O_2 nor CO_2 binds as tightly to hemoglobin as the CO.

21. The antidote for cyanide poisoning is thiosulfate, $S_2O_3^{2-}$. Thiosulfate ions react with CN^- to form thiocyanate ions (SCN^-).

$$CN^-(aq) + S_2O_3^{2-}(aq) \longrightarrow SCN^-(aq) + SO_3^{2-}(aq)$$

23. Process in which a molecule uses two or more sites to bond to a metal atom or ion. The chelating agent ties up the metal to keep it from disrupting metabolic processes. Two chelating agents are BAL

$$\text{H}\cdot\text{OCH}_2\text{-CH-CH}_2\text{-S}^{\text{-H}}$$

and EDTA

25. Amygladin is associated with cyanide, CN^-. Amygladin reacts with water in acidic conditions, hydrolyzes, to produce HCN.

27. a. Acute poisoning is poisoning that occurs with an immediate, severe response.
    b. Subacute poisoning is a response or set of symptoms resulting from poisoning that is at a lower dose such that an acute response is not triggered.
    c. Chronic poisoning is a condition that results from prolonged and continuous exposure to a poison at doses below the lethal level.

29. British Anti-Lewisite.

$$\text{H}\cdot\text{OCH}_2\text{-CH-CH}_2\text{-S}^{\text{-H}}$$

The -SH groups are the active sites. The structure for BAL tied to a heavy metal is

31. Lead levels are usually given in micrograms per deciliter of blood, µg/dL.

33. The solder used in joints in copper pipe contains lead. This lead will dissolve in acidic drinking water.

35. Acetylcholinesterase breaks down acetylcholine in a stimulated synapse. This releases choline at a receptor. This allows the synapse relax or rest and be ready to fire again to send another impulse to the receptor site. Anticholinesterase poisons deactivate acetylcholinesterase and prevent the breakdown of acetylcholine. The continued presence of acetylcholine keeps the synapse firing and over stimulates the nerve. This will continue until new acetylcholinesterase is produced.

acetylcholine

choline

37. Parathion, malathion and carbaryl are all anticholinesterase poisons.

39. The compound could interfere with fetal development. It would be a teratogen

41. The Ames test provides a quick way to determine whether a compound is a mutagenic hazard. It also is a good indicator of the carcinogenicity of a substance because most mutagens are also carcinogens. The time period for a test is only a few days instead of the longer periods associated with life cycles of other species. The test uses a population of about 100 million bacteria that cannot synthesize the amino acid histidine. This population is added to a mixture of agar suspension and the compound to be tested. The mixture is added to a culture dish and incubated for several days. If the compound being tested is a mutagen some bacteria will start to produce histidine.

282

43. The initiation step of carcinogenesis requires the alteration of a cell's DNA by a chemical, viral, high energy (radiation) or physical agent. This alteration results in a new generation of cells that do not have the normal DNA sequence. The change in DNA may be so great that abnormal cell function and growth may result.

45. Exposure of an organism to the MTD, maximum tolerated dose, for a prolonged period amounts to sustained or chronic wounding. This serves as a promoter of cancer and increases the number of cancers. Compounds may appear to be carcinogenic because test conditions are wrong.

47. Heavy metals typically are at the bottom of a group or in the transition series. Arsenic is a metalloid but it is classified with the heavy metals because it reacts with sulfhydryl groups. Some elements classified as heavy metals are shaded below.

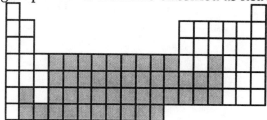

Solutions for Selected Problems

1. The question asks for the total lethal dose. The lethal dose is given as 50 mg / kg body weight. This means the person's body weight of 150 lb must be converted to kilograms. The lethal dose is 3400 mg.

$$\frac{150 \text{ lb}}{1} \times \frac{.454 \text{ kg}}{1 \text{ lb}} \times \frac{50 \text{ mg}}{1 \text{ kg}} = 3400 \text{ mg}$$

3. The question asks for the total dose to be fatal for 50% of a population of people with a body mass of 75 kg. The LD_{50} of sodium benzoate for rats is 4 g/ kg. If the rat data is assumed to apply to humans the required dose is 300 g.

$$\frac{75 \text{ kg}}{1} \times \frac{4 \text{ mg}}{1 \text{ kg}} = 300 \text{ g sodium benzoate}$$

The number of cans of soft drink (soda) that contain this much sodium benzoate is 845 cans. You can see this is a lot of soda.

$$\frac{1 \text{ can soda}}{355 \text{ g soda}} \times \frac{1.0 \text{ g soda}}{0.001 \text{ g sodium benzoate}} \times \frac{300 \text{ g sodium benzoate}}{LD_{50} \text{ dose}} = \frac{845 \text{ cans}}{LD_{50} \text{ dose}}$$

Speaking of Chemistry

Name _____

Toxic Substances

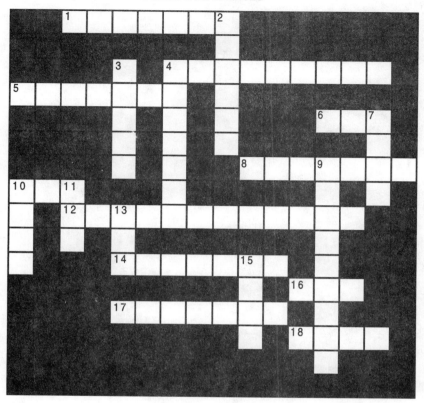

Across

1 Heavy metal poison able to react with sulfhydryl (-SH) groups in enzymes.
4 Substance that can cause genetic mutations.
5 Also called a "sequestrant".
6 A polymeric molecule that carries the genetic code in its structure.
8 EDTA is used to remove Pb by forming a _____ ion.
10 Abbreviation for the concentration measure, parts per billion.
12 Cyanide ion and carbon monoxide are classed as_____ poisons.
14 Chemicals like caffeine that can alter the structure of DNA.
16 Abbreviation for maximum tolerated dose.
17 Ion, NO_2^{1-} , converted to potential mutagen, HNO_2, nitrous acid.
18 Chimney _____ is a carcinogen.

Down

2 Contain the natural carcinogen D-Limonene.
3 Anticholinesterase poison developed during World War II.
4 Heavy metal poison.
7 Developed method to test for chemical mutagens and carcinogens using bacteria.
9 Second stage in the carcinogenesis process.
10 Peculiar craving for chalk, paint flakes, clay.
11 Abbreviation for British Anti-Lewisite antidote for heavy metal poisoning.
13 Abbreviation for concentration measure equal to 1 mg per kg.
15 Abbreviation for the calcium salt used to treat lead poisoning.

Bridging the Gap   I

Internet and the
United States Food and Drug Administration, FDA

The Center for Food Safety and Applied Nutrition, CFSAN

This division of the FDA deals with food contaminants including poisons. The FDA has a specific page that provides information about poisons in foods. This web page can be reached by using the following universal resource locator (URL). The FDA will update these pages so their exact appearance will change.

http://vm.cfsan.fda.gov/~lrd/pestadd.html

When you access this web site you will see a number of underlined subject headings. The information displayed on the page as of May 1996 is shown below. Every one of the underlined lines of text is interactive. The interactive topics can be selected by clicking on the text. Each one of these active lines leads to another page with a different URL address. Typically the content under any of these headings can be down loaded to a disk in the form of text or it can be printed. The Internet page is maintained by the FDA Center for Food Safety & Applied Nutrition (CFSAN). Updated versions will look slightly different.

Pesticides & Chemical Contaminants

Background

- Lead Threat Lessens, But Mugs can Pose Problem
- Mercury in Fish: Cause for Concern?
- Monosodium Glutamate-MSG-
- Pesticides: FDA Reports on Pesticides in Food (FDA Consumer 6/93)
- Pesticides: FDA 1993 Pesticide Program Residue Monitoring Report (10/94)
- Pesticides: FDA 1994 Pesticide Program Residue Monitoring Report (10/95)
- Pesticide Analytical Behavior Data (Pestrak and Pestdata files)

Guidance Documents for Industry
- Action Levels for Poisonous or Deleterious Substances in Human Food and Animal Feed.

For content Questions send E-Mail to oco@fdacf.ssw.dhhs.gov
Go BACK to the CFSAN/FDA food and consumer information page

This page last updated by lrd@vm.cfsan.fda.gov on 05/06/96

Your assignment is to open the topic " •Lead Threat Lessens, But Mugs can Pose Problem".

Once you have opened this category, read what the FDA says about the health risks associated with using ceramic mugs and containers. Record your findings on the Concept Report Sheet.

You should open other categories that interest you and record your opinion of the FDA material. Record the URL for the topic you selected and give your reasons for the choice

Bridging the Gap   I

Name _____

Internet and the FDA
Pesticides & Chemical Contaminants
Concept Report Sheet

1.   The FDA page on Pesticides & Chemical Contaminants has the following URL

        http://vm.cfsan.fda.gov/~lrd/pestadd.html

    Give the universal resource locator (URL) for the FDA page that deals with possible lead poisoning. The title for the page is "Lead Threat Lessens, But Mugs can Pose Problem."

        http://_____

    What does the FDA say about the possibility of lead poisoning from ceramic objects that are improperly glazed?

    How did tested items from the United States compare with ones from other countries?

    What suggestions does the FDA give citizens to protect themselves from lead poisoning from improperly glazed ceramics?

2.   Give the universal resource locator (URL) for the FDA page that deals with mercury poisoning from fish. The title for the page is " •Mercury in Fish: Cause for Concern?"

        http://_____

    What does the FDA say about the hazard posed by mercury contamination in fish?

3.   What is the URL for one additional topic or category that interests you?

        http://_____

    Why did you select this topic?

Bridging the Gap   II   Name_____
Public Policy and a Safe Food:
A Case Study "Demolition Man"

Research seems to indicate that specific types of cancers are related to genetics, lifestyle, diet, and smoking. The cancer and health risks may be reduced by changing diet. A great financial burden is placed society and on the health care system because people consume unhealthy foods. Today the Food and Drug Administration has the duty to protect the safety of our food supply. This has been interpreted in a narrow traditional sense. The traditional attitude aims at testing food for spoilage, contamination by toxic substances, etc. Future technology will enable us to do laboratory tests for natural carcinogens, contaminant carcinogens, and other "harmful" substances.

Should the government take a hand in testing food and limiting the availability of food that has the potential to cause cancer, heart disease, and other health problems? What do you think about a proposal to outlaw the sale and distribution of foods that contain proven natural carcinogens? Foods like coffee that contain carcinogenic caffeine and methylglyoxal would be banned. Citrus fruits that contain the natural carcinogen D-Limonene also would be banned. Only healthy foods would be legal.

This premise was part of the social condition portrayed in the movie, "Demolition Man" starring Sylvester Stallone, Sandra Bullock, and Wesley Snipes. Demolition Man is set in the 21st century and the society in the United States has changed dramatically.

Computers, video systems and sensors are everywhere. These devices are used to monitor personal behavior and preserve a tranquil, "safe" society. Government is depicted as protecting the public health by banning unsafe foods like chocolate, potato chips, hamburgers, and pizza.

Your assignment has two parts. One is to view part or all of "Demolition Man" and develop arguments concerning the limits of government action dealing with "health" issues. One argument should support the current policy. The other should be an argument in favor of a more intrusive and protective health policy. It is less expensive and more fun to view the movie with a study group. Write your arguments on the Concept Report Sheet.

The second part of your assignment is to write your opinion about the kind of governmental involvement depicted in the film. Do you think it could ever happen? Why? Why not?

Specifics about the movie.
Produced by: Silver Pictures / Warner Brothers
Released in USA in 1993
Color by: Technicolor
Rating "Demolition Man" has a USA "R" rating

289

Bridging the Gap   II   Name_____

Concept Report Sheet
Public Policy and a Safe Diet: A Case Study "Demolition Man"

1.   Argument in favor of the present policy.

2   Argument in favor of a more protective health policy.

3.   What is your opinion about the kind of governmental regulation depicted in the film? Do you think it could ever happen in the United States? Why? Why not?

Solution to Speaking of Chemistry crossword puzzle
Toxic Substances

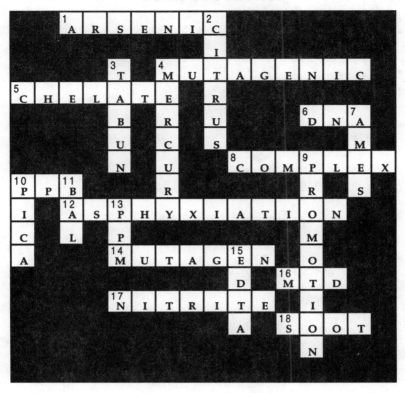

19 Chemistry and Medicine

19.1 Medicines, Prescription Drugs, and Diseases--The Top Tens

Over-the-counter and prescription drugs are defined. The top ten prescription drugs in the first quarter of 1994 are tabulated. The meanings of the trade name (brand name) and generic name for prescription drugs are explained. The change in the types of drugs prescribed and the maladies treated is traced from 1900 to 1993. The leading causes of death in 1993 are tabulated in Figure 19.1

Objectives
After studying this section a student should be able to:
give the definition for over-the-counter drug
give the definition for prescription drug
name the condition for which most prescriptions are written
identify the classes of drugs that are most prescribed

Key terms
over-the-counter drugs   prescription drug   generic name
systematic name   U.S. Food and Drug Administration   trade name

19.2 Drugs for Infectious Diseases

The history of chemotherapy dating back to Paul Erlich in 1904 is summarized. The structures of sulfa drugs are given. The action of sulfa drugs on bacteria is illustrated and explained. Antibiotics and pathogenic bacteria are defined. A brief history of the discovery and use of penicillins is given. The structures and modes of action for penicillins, tetracylines and cephalosporins are described.

Sulfanilamide, a sulfa drug

Penicillin G

Cefaclor, a cephalosporin

Tetracycline

Objectives
After studying this section a student should be able to:
give a definition for chemotherapy, sulfa drug, antibiotic, penicillins, cephalosporins, tetracylines
explain what is meant by a prodrug
describe how sulfa drugs like sulfanilamide work to stop bacteria growth
explain how all penicillins function to kill growing bacteria
tell how cephalosporins and tetracyclines control bacteria
tell what it means when we say a bacteria is "antibiotic resistant"
explain why "resistant" bacteria are a public health problem

Key terms

chemotherapy	infectious diseases	sulfa drugs
prodrugs	folic acid	para-aminobenzoic acid
antibiotics	antimicrobial	pathogen
penicillins	penicillin G	cell wall development
tetracyclines	cephalosporin	protein synthesis
rifampin	RNA synthesis	antibiotic resistance

19.3   AIDS, A Viral Disease

The history and scope of the spread of acquired immune deficiency syndrome (AIDS, caused by the virus HIV) is discussed. The steps in the process of HIV invasion of a T lymphocyte are illustrated. A model for the action of AZT in retarding the progression of AIDS is described.

Objectives

After studying this section a student should be able to:

give the definition for HIV

explain what is meant by a retrovirus

describe the process in which an HIV virus attacks a T lymphocyte and causes the production of a new virus

tell how AZT alters the HIV synthesis of its DNA

name the base that AZT is believed to replace in DNA

explain why AZT is not a cure for AIDS

identify two problems that investigators face when doing AIDS research

Key terms

acquired immune deficiency syndrome	AIDS
Human immunodeficiency virus, HIV	retrovirus
RNA-directed synthesis of DNA	DNA
DNA-directed synthesis of RNA	reverse transcriptase
T cells	AZT, azidothymidine

19.4   Steroid Hormones

Hormones and their interaction with receptors are defined. The structures for natural and synthetic female sex hormones are illustrated. The most commonly used hormones in oral contraceptives are identified.

estradiol   and ethynyl estradiol   Reasons for prescribing Premarin for women after menopause are given. The controversial drug RU-486 is described.

Objectives

After studying this section a student should be able to:

give the definition for hormone

give the definition for a receptor

sketch the shape of the basic steroid carbon skeleton

tell what development made the pill possible

name the two most common synthetic hormones used in oral contraceptives

tell what RU-486 is prescribed for in France, Sweden and the United Kingdom

294

explain why Premarin is prescribed
distinguish between the structures for estradiol and ethynyl estradiol

Key terms

steroid hormones
chemical messengers
progesterone-like

hormones
oral contraceptives
RU-486, Mifepristone

receptor
estrogen-like
Premarin

19.5   Neurotransmitters

Neurotransmitters are defined. Regulating functions of serotonin and norepinephrine are described. Physiological effects of the combination of both compounds are given. Structures for both compounds are illustrated. The roles of both compoundsin the function of nerve synapses are summarized. The common monoamine structural group, $-NH_2$, is identified in both neurotransmittors and antidepressants. The blood-brain barrier is described and its protective function is explained. Parkinson's disease and its treatment with the prodrug L-dopa are described. The link between excess dopamine and schizophrenia is given. Three types of antidepressant drugs are described in terms of their function: tricyclic antidepressants (Elavil), monoamine oxidase inhibitors , those that prevent serotonin recapture by the original neuron, (Prozac). The use of epinephrine to treat acute asthma attacks and anaphylactic shock from acute allergy attacks is discussed.

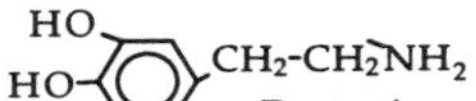

Dopamine 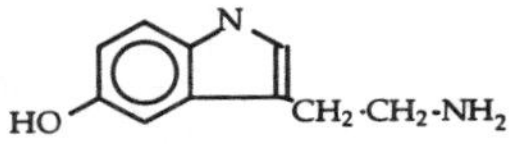   Serotonin 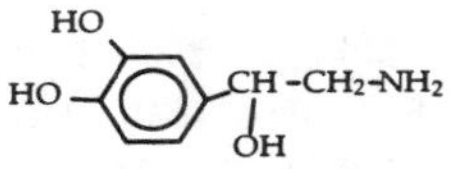   Norepinephrine

Objectives

After studying this section a student should be able to:

tell what body functions norepinephrine controls
tell what body functions serotonin regulates
tell what activity serotonin and norepinephrine appear to control together
summarize the steps in the cycle of impulse transmission in a nerve synapse
tell what structural feature all neurotransmitters have in common
describe the benefits that the blood-brain barrier provides
tell why L-dopa instead of dopamine is used to treat Parkinson's disease
tell what types of molecules can cross the blood-brain barrier
describe the physiological effects that epinephrine produces
name the conditions that epinephrine is used to treat

Key terms

neurotransmitters
monoamines
monoamine oxidase inhibitors
epinephrine
Parkinson's disease
serotonin concentration

serotonin
amino group
Prozac
dopamine
schizophrenia
allergic reactions

norepinephrine
monoamine oxidase
adrenalin
L-dopa
tricyclic antidepressants
anaphylactic shock

19.6   Painkillers of All Kinds

Analgesics are defined. Three classes of analgesics- -over-the-counter-, prescription-, and illegal- are described. Alkaloids are defined. The opium poppy is identified as the main source of opium which is a mixture of at least 20 different alkaloids. Morphine is named as the most abundant alkaloid in opium. Opoids are defined as compounds with morphine like properties. Mild analagesics are described. Aspirin's structuresynthesis and properties are discussed The link between aspirin and Reye's Syndrome is described.

Mild analagesics are described. Aspirin's structure , synthesis and properties are discussed. The link between aspirin and Reye's Syndrome is described. Structures and properties of other over-the-counter antipyretic and anti-inflammatory agents such as acetaminophen (Tylenol) and Ibuprofen are given.

Objectives

After studying this section a student should be able to:

state the definition for alkaloids

tell what compound is primarily responsible for the effects of opium

name two compounds derived from morphine

give the definitions for analgesic, antipyretic, and anti-inflammatory

explain why aspirin can cause stomach bleeding

tell why aspirin tablets develop a vinegar-like odor

describe the link between Reye's syndrome and aspirin

name three nonaspirin pain relievers

tell which analgesics contain a carboxylic acid group

Key terms

analgesics	alkaloids	aspirin
heroin, diacetate ester of morphine	morphine	anti-inflammatory
codeine, methyl ether of morphine	opium	Reye's syndrome
mild analgesics	Demerol, meperidine	naproxen, Aleve
acetaminophen, Tylenol	buffered aspirin	antipyretic
Ibuprofen, Advil and Nuprin	opoid	

19.7   Mood-Altering Drugs, Legal and Not

The effect of depressants on the nervous system is described. Structures and the mode of action for barbiturates, phenobarbital, and benzodiazeprines (Valium) are described. Stimulants such as amphetamines, cocaine, and crack are illustrated and described. Natural sources of hallucinogens like LSD, marijuana and mescaline are discussed. The structure and effects of PCP, "angel dust", are described. The DEA drug classification system is described and listed in Table 19.2.

Objectives

After studying this section a student should be able to:

state the definitions for stimulants, depressants, and hallucinogens

give an example of each of the following: a stimulant, a depressant, an hallucinogen

tell what is meant by a scheduled substance

name a schedule 1 substance and name its applications

name a schedule 2 substance and name its applications

name a schedule 3 substance and name its applications

tell what GABA does in the nerve synapse

explain how crack is prepared

describe how depressants like barbiturates act to inhibit nerve firing and prolong the binding of GABA to its receptors

explain why a combination of alcohol and barbiturates is more dangerous than each is separately

name the natural sources of lysergic acid diethylamide, LSD

name the natural source of marijuana and identify the active agent in marijuana

tell what amphetamines are, why they are restricted, and identify previous applications

describe the physiological effects of the following: THC, alcohol, LSD, PCP, crack

Key terms

USDEA	mood-altering drugs	depressants
Schedule 1 drugs	Schedule 2 drugs	insomnia
gamma-aminobutyric acid, GABA	sedation	benzdiazepines
angel dust, PCP, phenylcyclidine	methamphetamine	anxiety
tranquilizers	barbiturates	ethanol
mescaline	Librium and Valium	amphetamines
lysergic acid diethylamide, LSD	stimulants	caffeine
tetrahydrocannabinol, THC	cocaine	narcolepsy
crack	marijuana	hallucinogen

19.8   Colds, Allergies, and Other "Over-the-Counter" Conditions

The connection between allergic reactions, hay fever, and histamines is described. The structures of histamine and antihistamines are given. Decongestants, antitussives, and expectorants are defined. A summary of recommended guidelines for selecting OTC products completes the section.

Objectives

After studying this section a student should be able to:

give definitions for histamines, antihistamines, decongestants, expectorants

tell why it is a good idea to know the physiological effects an over the counter drug

give an example for each of the following:

decongestants, antitussives, expectorants, analgesics, antihistamines

state the four recommended guidelines for selecting OTC products

Key terms

OTC, over-the-counter	decongestants	antihistamines
histamine	allergic symptoms	antitussives
expectorants		

19.9   Heart Disease and Its Treatment

Cardiovascular disease is described. The link between heart disease and atherosclerorsis is explained. Hypertension is defined. Angina, ischemia and arrhythmias are described. The purpose and function of cholesterol-lowering drugs, diuretics, vasodilators, beta blockers, calcium channel blockers, tissue plasinogen, and aspirin are explained. Structures for representative compounds like nitroglycerine for angina are given.

$$\text{Nitroglycerine} \quad \begin{array}{l} H_2C\!-\!ONO_2 \\ CH\!\cdot\!ONO_2 \\ H_2C\!-\!ONO_2 \end{array}$$

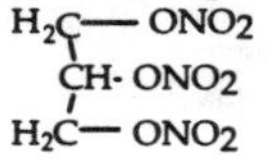

Objectives

After studying this section a student should be able to:

    give the definition for cardiovascular disease
    describe the symptoms for angina
    tell what is meant by the term "heart failure"
    explain why cholesterol-lowering drugs are prescribed and name an example
    state the function of a diuretic and give an example
    name a vasodilator and tell why it is prescribed
    state the purpose of beta blockers
    tell why calcium channel blockers are prescribed
    describe how calcium channel blockers work
    name three clot dissolving drugs used to treat heart attack victims
    describe how regular doses of aspirin are believed to help prevent heart attack

Key terms

cardiovascular disease	plaque	atherosclerosis
high blood pressure	hypertension	diuretics
ischemia	cardiac arrhythmias	angina
cholesterol-lowering drugs	blood cholesterol	blood volume
thiazide	vasodilators	nitroglycerine
NO	amyl nitrate	beta blockers
propranol, Inderal	beta-1 receptors	beta-2 receptors
calcium channel blockers	clot dissolving drugs	plasminogen
plasmin	streptokinase	urokinase
aspirin	prostaglandins	tissue plasminogen activator, TPA

19.10   Cancer and Anticancer Drugs

A general description of cancer is given. A summary of cancer incidence for 11 different body sites is illustrated graphically in Figure 19.10. Three ways to treat cancers are listed. Two types of cancer chemotherapy are described. The function of alkylating agents to interfere with DNA replication is explained. Antimetabolite drugs are described. The reason for using synergistic combinations of drugs is explained.

298

Objectives
After studying this section a student should be able to:

list three ways that cancers are treated

describe how survival of cancer patients has changed between 1930 and 1990

name the cancer that has the highest % increase in incidence between 1971 to 1991

name the two types of cancer chemotherapy drugs

tell why alkylating agents are used to treat cancer even though they attack normal cells

explain why combination chemotherapy is now used instead of single drugs

Key terms

cancer	DNA damage	physical
biological	chemical agents	treatment
surgery	irradiation	chemicals
cancer incidence	cancer death rates	mustard gas
nitrogen mustards	alkylating agents	alkyl groups
antimetabolites		

Suggested Additional Readings and Resources

Aharonowitz, Yair and Gerald Cohen. "The Microbiological Production Of Pharmaceuticals" Scientific American Sept. 1981: 40.

Ashby, J. and H.Tinwell. "Is Thalidomide Mutagenic?" Nature June 8, 1995: 453.

Cohen, Jon. "Bringing AZT to Poor Countries" Science Aug. 4, 1995: 624.

Gutfield, Greg. "Penicillin's No Villain: You May Not be Allergic to it After All" Prevention Jan. 1992: 18.

Holloway, Marguerite. "Rx For Addiction" Scientific American Mar. 1991: 94.

Kartner, Norbert and Victor Ling. "Multidrug Resistance In Cancer " Scientific American Mar. 1989: 44.

Lawn, Richard M. "Lipoprotein(a) In Heart Disease" Scientific American June 1992: 54.

Lipkin, Richard. "Tamoxifen Puts Cancer on Starvation Diet" Science News Nov. 5, 1994: 292. 182.

Liotta, Lance A. "Cancer Cell Invasion and Metastasis" Scientific American Feb. 1992: 54.

Melzack, Ronald. "The Tragedy Of Needless Pain" Scientific American Feb. 1990: 27.

Musto, David F. "Opium, Cocaine And Marijuana In American History" Scientific American July 1991: 40.

Oppenheim, Mike. " 7 Good Drugs You Should Know About" Family Circle June 29, 1993: 59.

Raloff, Janet. "This Fat May Fight Cancer Several Ways" Science News Mar. 19, 1994: 182.

Rose, Anthony H. "New Penicillins" <u>Scientific American</u> Mar. 1961: 66.

Rozin, Skip. "Steroids: a Spreading Peril" (includes related article on steroid use by Chinese swimmers) <u>Business Week</u> June 19, 1995: 138.

Tarpy, Cliff. and José Azel. "Straight: A Gloves-off Treatment Program" <u>National Geographic</u> Jan. 1989: 48.

Ulmann, Andre et al. "Ru-486" <u>Scientific American</u> June 1990: 42.

White, Peter T. and José Azel. "Coca--An Ancient Herb Turns Deadly" <u>National Geographic</u> Jan. 1989: 3.

Winter, Ruth. "Which Pain Relievers Work Best?" (includes related information on drug interactions) <u>Consumers Digest</u> Sept.-Oct. 1994: 76.

Answers to Odd Numbered Questions for Review and Thought

1. Malaria, pneumonia, bone infections, gonorrhea, gangrene, tuberculosis, and typhoid fever are bacterial diseases. Polio, AIDS, and Rubella (German measles) are viral diseases.

3. The United States Food and Drug Administration, FDA, has the responsibility for classifying drugs as either over-the-counter drugs or prescription drugs.

5. Three major classes of antibiotics are penicillins, cephalosporins, and tetracyclines.

7. Chemotherapy is the treatment of disease with chemical agents.

9. A retrovirus is a virus that uses RNA directed synthesis of DNA instead of the usual DNA-directed synthesis of RNA.

11. a. Vasodilators are used to treat heart disease and asthma. The vasodilator relaxes the walls of blood vessels and creates a wider passage for blood flow. This lowers blood pressure and reduces the amount of work that the heart must do to pump blood. It also enables a person to breathe more deeply and easily by dilating the bronchial tubes.
    b. Alkylating agents are used to treat cancer. Alkylating agents react with the nitrogen bases of DNA in cancer cells and normal cells. Alkyl groups are added to the nitrogens in the bases. This has an effect on both cancer cells and normal cells, but cancer cells are usually dividing and duplicating DNA more often, so the cancer cells are impacted more.

300

c. Beta blockers are used to treat heart disease. Propranolol (Inderol) is used to treat angina, cardiac arrhythmia, and hypertension. Beta blockers act to keep epinephrine and norepinephrine from stimulating the heart.

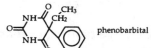

Inderal, Propranolol

d. Antimetabolites are used to treat cancer. Antimetabolites interfere with DNA synthesis, and cancer cells are more susceptible than normal cells because the cancer cells are generally replicating DNA more frequently than normal cells.

13. Histamines and neurotransmitters bind to receptor sites.

15. a. Lovastatin is a cholesterol lowering drug.
    b. Methotrexate is an antimetabolite used to treat cancer.
    c. Chlorphenirimine (in Chlortrimeton) is an antihistamine
    d. Pseudoephedrine is a decongestant.

17. Codeine is
    a. an analgesic   b. an opoid   d. a controlled substance, a Schedule 2 drug.

19. a. Angina results from heart disease with the symptom of chest pain on exertion.
    b. Arrhythmia is a heart disease; its symptom is an abnormal heart rhythm.

21. Prostaglandins are lipids derived from 20-carbon carboxylic acids. They control blood clot formation, blood lipid levels, blood vessel contractions, nerve impulses, and inflammation response to injury and infection.

23. A barbiturate such as phenobarbital, is a depressant. Barbiturates bind to a receptor for GABA. This keeps channels for chloride ion, Cl^{1-}, transmission open. This inhibits transmission of nerve impulses. The physiological effects are a progression from sedation or relaxation, to sleep, to general anesthesia, to coma and death.

25. The mustard alkyl groups attach to the nitrogen bases in DNA, often guanine. This bonding physically blocks base pairing, prevents DNA replication and stops cell division.

27. Serotonin, dopamine and norepinepherine are neurotransmitters.

29. Premarin is a female hormone replacement. It slows bone loss that normally occurs with age.

31. a. Tylenol is an analgesic.
    b. Ibuprofen is both an analgesic and an antiinflammatory.
    c. Naproxen is both an analgesic and an antiinflammatory.

33. Cocaine is derived from the leaves of the coca plant. Usually cocaine is distributed as a white crystalline solid. It is the hydrochloride salt of the parent molecule. Crack is a nonsalt form of cocaine produced from a water solution of the hydrochloride salt that is

treated with baking soda, $NaHCO_3$. Chunks form when the water is evaporated off. These pieces of solid are a mixture of cocaine and sodium bicarbonate. The solid is called crack because it "pops" when heated and the sodium bicarbonate decomposes to release CO_2.

35. a. Antihistamines block histamine receptors and are used to treat allergic reactions.
    b. Analgesics are used to diminish pain.
    c. Decongestants are used to shrink nasal passages.
    d. Antitussives, also known as cough medicines, are used to control cough.
    e. Expectorants are used to loosen fluids and allow cough to clear respiratory tract.

37. Compounds that have a mustard like odor and a general structure like this

and a specific mustard is . Mustards were used as poison gases during World War I. Now they are used in chemotherapy to treat cancer.

39. The R group in penicillin G is enclosed in the box in the formula below.

Some penicillins are effective only against certain bacteria. Some bacteria are resistant to particular penicillins.

41. Hallucinogens cause a person to experience vivid illusions, fantasies, and hallucinations. Examples of hallucinogens are mescaline, PCP, and LSD.

43. a. Morphine is a more effective pain killer than heroin and codeine.
    b. Heroin is not a natural alkaloid.
    c. Heroin is so addictive that it is not legal to sell or use it in the United States.

45. Nitroglycerine is a heart muscle relaxant. It is used to treat angina which is a symptom of heart disease.

47. The blood-brain boundary is defined physically by small openings in capillaries and the astrocyte cell membranes that prevent passage of large polar molecules. The barrier keeps large toxic polar molecules from passing into the brain. Drugs must be soluble in blood and soluble in the lipid layer of the membrane in order to reach the brain.

49. Both questions 48 and 49 use Figure 19.9, not Figure 19.7.
    a. +80 %
    b. +25 %, +20 %
    c. +35 %, + 40 %

Speaking of Chemistry

Name _____

Chemistry and Medicine

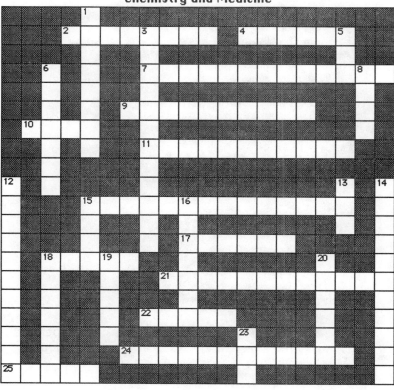

Across

2 Makes up about 10% of crude opium.

4 Name for chest pain on exertion.
7 Schedule 3 controlled substance.
9 Abnormal heart rhythm.
10 Analgesics are used to treat _____.
11 Also known as adrenalin.
15 Prozac; prevents recapture of serotonin.
17 A Schedule 1 controlled substance.
18 Sulfanilamide is a_____ drug.
21 Name for fever reducers.
22 Sulfa drugs shut off _____ acid production.
24 LSD, PCP and tetrahydrocannabinol.
25 _____ syndrome is associated with aspirin treatment of children with flu and chicken pox.

Down

1 A neurotransmitter, HO-HO-⟨○⟩-$CH_2 \cdot CH_2 \cdot NH_2$
3 High blood pressure.
5 Azidothymidine, used to treat AIDS.
6 Stimulant derived from coca plant leaves.
8 Acquired immunodeficiency syndrome, caused by the virus HIV.
12 Nitroglycerine is a _____ .
13 Drugs available to anyone.
14 Animal steroid associated with heart disease.
15 _____ nitrate is used to treat angina.
16 Synthetic estrogen is _____ estradiol.
18 _____, irradiation and chemicals are the three therapies used to treat cancer.
19 Excessive prostaglandin causes _____.
20 The third leading cause of death for all ages in 1993.
23 Human immunodeficiency virus.

Bridging the Gap
Concept Report Sheet

Name_____

FDA Prescription Drug Evaluation and Drug Prices

This is a two part exercise. The first part deals with the time required to bring a new drug to market. The second deals with prescription drug prices.

1.   FDA Prescription Drug Evaluation: Too Slow or Just Right

The United States Food and Drug Administration has the responsibility for reviewing and approving drugs before they can be sold in the USA. The FDA must decide on the safety and effectiveness of each new drug. This process can take as long as 12 years from the initial discovery to final approval and public sale. Many countries like France and Great Britain have equally thorough drug evaluation procedures that are much quicker.

Very often a potential drug is discovered by a U.S. pharmaceutical company. The company starts the evaluation process simultaneously in the United States and overseas. The testing process is completed overseas and the drug is approved for sale, but the FDA is still doing its review. The drug is typically unchanged when the FDA approves it for use in the USA a few years later. In the mean time the drug is not available to the public in the USA.

Your assignment is to discuss this situation with your classmates, family or friends. You are supposed to develop two questions you would want answered about the difference in evaluation times. Write your two questions in the space below.

1.

2.

2.   Prescription Drug Prices

Prescription drugs can be very expensive. There are trade name prescription drugs and generic name prescription drugs. The two are typically very different in price; the generics usually cost less.

Your assignment has two parts. The first part is to create a question you would ask a pharmacist or health care professional about the differences between generic and trade name prescription drugs. Then you should talk to a health care professional and actually ask your question. Record both your question and a summary of your conversationwith the health care professional. The second part is to discuss this situation with your study group, classmates, family or friends. Based on these discussions develop an opinion about why the generics cost less than the trade names.

1.   Question for health care professional

2.   Summary of conversation with health care professional

3.   Summary of your discussions and your opinion

Solution to Speaking of Chemistry crossword puzzle

Chemistry and Medicine

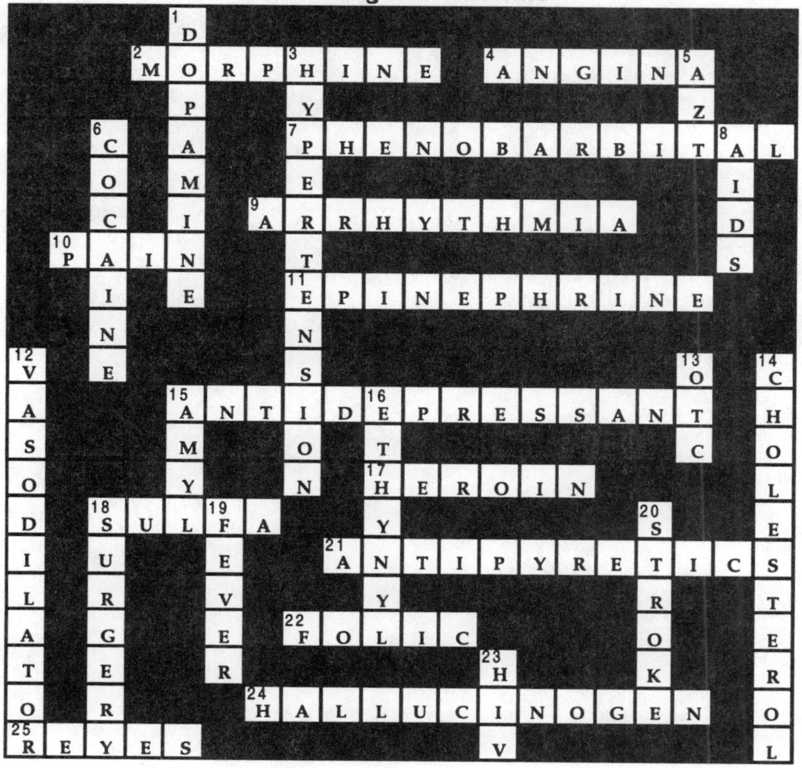
1 DOPAMINE
2 MORPHINE
3 HYPERTENSION
4 ANGINA
5 AZT
6 COCAINE
7 PHENOBARBITAL
8 AIDS
9 ARRHYTHMIA
10 PAIN
11 EPINEPHRINE
12 VASODILATOR
13 OTC
14 CHOLESTEROL
15 AMY
16 DETHYLY
17 HEROIN
18 SURGEER
19 FEVER
20 SROK
21 ANTIPYRETICS
22 FOLIC
23 HIV
24 HALLUCINOGEN
25 REYES
SULFA

20 Water--Plenty of It, but of What Quality?

20.1 How Can There Be a Shortage of Something as Abundant as Water?

The responsibility of the U. S. Environmental Protection Agency (EPA) to enforce federal laws regulating waste water and drinking water quality is described. Water is identified as the most abundant compound on the earth's surface. The small amount (2.5%) of fresh water and its uneven distribution are discussed. Surface water and groundwater are defined and named as the sources of usable water. Aquifers and artesian wells are described. The Ogallala aquifer in the region between South Dakota and Texas is discussed. Salt water intrusion into coastal fresh water supplies to produce "brackish" water is explained. Ground subsidence and sinkhole formation when aquifers are tapped by deep wells are discussed.

Objectives
After studying this section a student should be able to:
- give definitions for surface water, groundwater, aquifers, artesian wells
- identify the sources of surface water
- describe the problem that deep wells pose
- give the approximate depth of a "deep" well
- explain what is meant by aquifer depletion
- describe the relationship between aquifer depletion and salt water intrusion
- explain how sinkholes are created by pumping water from aquifers

Key terms
surface water	groundwater	aquifers
artesian wells	shallow well	deep well
aquifer depletion	brackish water	sinkholes

20.2 Water Use and Reuse

The major consumers of fresh water are tabulated. Agriculture is identified as the largest single user of fresh water. Industrial use of cooling water is explained. Potable water is defined. Groundwater and surface water are named as the source of potable water. The average water usage per person per day for bathing, laundering drinking, etc., is tabulated. Average individual drinking water consumption is estimated as 2 gallons per day. Natural recycling of water molecules is discussed. Aquifer depletion is described. The replacement of aquifer water by groundwater recharge is explained.

Objectives
After studying this section a student should be able to:
- name the major users of fresh water in the United States
- describe why water is used as a coolant
- give definitions for thermal pollution, potable water, dual water system
- explain why thermal pollution creates problems for aquatic life
- explain why dual water systems are receiving attention
- describe the purpose of groundwater recharge

Key terms

industrial water   thermal pollution   potable
dual water systems   wastewater   sewage
groundwater recharge

20.3   What is the Difference Between Clean Water and Polluted Water?

Pollution is defined. The official U.S. Public Health Service classes of pollutants are identified and tabulated. Pollutants are identified as: heat, radioisotopes, toxic metals, acids, toxic anions, organic chemicals, infectious agents (pathogens), sediment from land erosion, plant nutrients, plant and animal matter. The Clean Water Act of 1977 is described. The discharger's responsibility to ensure the cleanliness of wastewater is explained. The role of the EPA is described.

Objectives

After studying this section a student should be able to:
give the definition for pollution
name the eight classes of pollutants identified by the U.S. Public Health Service
explain how The Clean Water Act of 1977 changed responsibility for keeping water clean
summarize the responsibilities of the EPA regarding water quality

Key terms

pollution   heat   radioisotopes
toxic metal   acids   organic molecules
alkalis   pathogens   "pure" natural water
human activity   natural processes   Clean Water Act of 1977
pollutants   wastewater effluent   Federal Water Pollution Control Act

20.4   The Impact of Hazardous Industrial Wastes on Water Quality

The problems that result from solid waste disposal in "old" style landfills are explained. The water pollution conditions that led to the Resource Conservation and Recovery Act of 1976 (RCRA) are described. The purpose of the "Superfund" to clean up hazardous waste disposal sites is described. The "cradle-to-grave" responsibility of waste generators is discussed. The RCRA requirement for hazardous waste "manifests" is explained. The concept of "secure" landfills is described. The prohibition against putting hazardous waste in landfills is cited as one reason for the incineration of hazardous waste. Legislation like California's Proposition 65 to prohibit the disposal of hazardous materials in the water supply is discussed.

Objectives

After studying this section a student should be able to:
describe the problems that "old landfills" pose
describe the situation that led to the creation of RCRA and the "Superfund"
describe the purpose of the "Superfund"
identify the agency that defines hazardous wastes
explain what the term "cradle-to-grave" means
tell the purpose of a "manifest" for hazardous wastes
state the three presently accepted ways to dispose of hazardous waste
describe some of the security features of a "secure landfill"
describe the purpose of laws like California's Proposition 65

Key terms

solid wastes	landfills	"responsible parties"
RCRA	Superfund	Resource Conservation and Recovery Act
cradle-to-grave	incineration	hazardous wastes
manifest	secure landfills	California Proposition 65

20.5   Household Wastes that Affect Water Quality

Common household products that become hazardous wastes are identified. The hazardous components of household wastes are classified as acids, caustics, organic solvents, etc. (See table 20.8) Problems faced when disposing of household wastes are explained. Methods for reducing the problem of hazardous household waste are discussed.

Objectives

After studying this section a student should be able to:

explain why the individual small amounts of household wastes can be a major source of pollution

explain why individual households have problems with the disposal of hazardous waste

tell how the amount of oil spilled by the Exxon Valdez compares with the amount of motor oil poured down the drain

Key terms

special disposal	disposal down a drain	trash disposal
household wastes	pesticide waste	hazardous household wastes

20.6   Measuring Water Pollution

Analytical chemistry is defined. Biological oxygen demand, BOD, is defined. The consumption of oxygen by organic waste is discussed. Concentration measures like parts per million (ppm) are explained. The solubility of oxygen in water is discussed. The changes in O_2 solubility with temperature are linked to thermal pollution. The method for calculating BOD is shown. The connection between BOD, dissolved oxygen, and fish kills is described and illustrated. The design of a general atomic absorption spectrometer (AA) is diagrammed. The use of the AA to measure ppm concentrations of metals is described. The general design of a gas chromatograph-mass spectrometer combination (GC/ MS) is illustrated. The operation of the gas chromatograph is described. Detection and measurement of organic waste using a GC /MS are described.

Objectives

After studying this section a student should be able to:

give the definition for analytical chemistry

state the definition for biological oxygen demand, BOD

name the types of products formed by oxidation of organic carbon

describe the relationship between BOD and dissolved oxygen

tell how BOD relates to survival of fish

tell how concentration of dissolved oxygen relates to water temperature

describe how an AA can be used to analyze water samples for dissolved metals

tell how a combination gas chromatograph-mass spectrometer operates

tell what kinds of pollutants are measured using a GC/MS

Key terms

analytical chemistry	dissolved oxygen	biochemical oxygen demand, BOD
metabolize	burning	dissolved organic matter
oxygen depletion	parts per million	solubility of oxygen
aerobic bacteria	anaerobic bacteria	spectrophotometry
original energy levels	higher energy levels	atomic absorption spectrophotometry, AA
mass spectrum	base peak	mass spectrometer, MS
positive ions	standard solution	gas chromatograph, GC
chromatogram	GC/MS	gas chromatograph-mass spectrometer

20.7 How Water Is Purified Naturally

The seven different natural water purification processes are identified and briefly described. Biodegradable and nonbiodegradable substances are defined. Branched-chain detergents and linear-chain detergents are used as examples of nonbiodegradable and biodegradable organics. Problems posed by hard chlorinated pesticides are discussed. The fat solubility of DDT and related compounds is discussed. The concentrating effect of the food chain on fat soluble polychlorinated hydrocarbons is described. The reason for banning DDT in the U.S. is explained.

Objectives

After studying this section a student should be able to:

> name seven different natural water purification processes
> state the definitions for biodegradable and nonbiodegradable
> describe how linear chain and branched chain detergents differ in biodegradability
> give the definition for a persistent insecticide
> tell why chlorinated hydrocarbons like DDT are hazardous
> tell how small concentrations of DDT in fresh water can lead to serious health hazards

Key terms

water cycle	distillation	crystallization
aeration	filtration	sedimentation
settling	oxidation	dilution
biodegradable	nonbiodegradable	natural purification
food chain	chlorinated hydrocarbon	polychlorinated hydrocarbon
fat soluble	detergent	branched chain detergent
DDT	linear-chain alkylbenzenesulfonate detergent	

20.8 Water Purification Processes: Classical and Modern

The old fashioned cess-pool method for sewage treatment is described. The evolution of sewage treatment plants is summarized. The levels of sewage treatment are defined: primary treatment, secondary treatment, tertiary treatment. The types of wastes removed by each type of treatment are identified. Special problem pollutants are named. The reasons for carbon black filtration and denitrifying bacteria are explained.

Objectives

After studying this section a student should be able to:

> describe the water purification processes used in primary wastewater treatment
> describe the water purification processes used in secondary wastewater treatment
> describe the water purification processes used in tertiary wastewater treatment
> tell what kind of wastewater treatment is required by the 1972 Clean Water Act

explain why filtration through carbon black is used in tertiary wastewater treatment
tell why denitrifying bacteria are used in wastewater treatment

Key terms

water purification	outhouses	cesspools
sedimentation	chlorinator	primary wastewater treatment
chlorination	aerobic	1972 Clean Water Act
anaerobic	sludge	secondary wastewater treatment
toxic metal ions	nitrate ions	tertiary wastewater treatment
ammonium ions	carbon black	denitrifying bacteria
adsorbed	adsorption	

20.9 Softening Hard Water

Hard water is defined. The metal ions Ca^{2+}, Mg^{2+}, Fe^{3+} and Mn^{2+} that make water "hard" are identified. The undesirable properties of hard water are identified. Methods for "softening" hard water are described in detail. The health risks associated with the lime-soda water softening process are explained.

Objectives

After studying this section a student should be able to:
state the definition for hard water
name the ions present in hard water
name three problems associated with hard water
describe how to soften hard water that contains Ca^{2+} and Mg^{2+} ions
describe how to soften hard water that contains Fe^{2+} and Mn^{2+} ions
tell what happens to soap when it dissolves in hard water

Key terms

hard water	hardness	lime-soda
calcium carbonate	calcium bicarbonate	magnesium hydroxide
magnesium bicarbonate	soft water	oxidation with air
aeration	pH	precipitate
acidic		

20.10 Chlorination and Ozone Treatment of Water

The effectiveness of chlorination of water in controlling water borne diseases like typhoid is described. The oxidizing power of chlorine is described as the reason for its disinfecting power. Disinfection byproducts are defined. The potential hazards of disinfection byproducts are explained. Disinfection of water with ozone is described. Bromate ion is identified as one of the ozonation disinfection by-products.

Objectives

After studying this section a student should be able to:
tell why gaseous chlorine and ozone are used to treat water
name the water-borne diseases chlorination is used to control
give the definition for disinfection byproducts and tell why they are a problem
describe the problems posed by ozonation and chlorination

Key terms

chlorination	chlorine	oxidizing
cholera	typhoid	water-borne diseases
paratyphoid	dysentery	disinfection byproducts
giardiasis	chloroform	dichloromethane
chlorobenzene	carcinogens	trichloroethylene
mutagenic	ozonation	ozone
	bromate ion	

20.11   Fresh Water from the Sea

The percentage of salts in ocean water is tabulated. The acceptable limit on dissolved salts is given as less than 500 ppm. Reverse osmosis and solar distillation are identified as water purification methods.

Objectives

After studying this section a student should be able to:

give the definitions for permeable membrane, semipermeable membrane, osmosis, reverse osmosis, osmotic pressure, distillation, still

tell why desalination is necessary

diagram an apparatus for producing fresh water from salt water using reverse osmosis

sketch a diagram of a simple solar still

Key terms

fresh water	ppm	solar distillation
reverse osmosis	permeable	semipermeable membrane
membrane	osmosis	osmotic pressure

20.12   Pure Drinking Water for the Home

The properties of bottled drinking water are described. Spring water is defined. The effectiveness of distillation, reverse osmosis and carbon filtration as purification methods is summarized.

Objectives

After studying this section a student should be able to:

name three commonly accepted methods for water purification

explain why it is important to read labels on bottled water

name the kinds of pollutants that pass through purification by reverse osmosis

name the kinds of pollutants that pass through purification by distillation

name the kinds of pollutants that pass through purification by carbon filtration

Key terms

bottled water	"spring" water	home water-treatment
trace pollutants	analysis on the label	

20.13   What About the Future?

Increasing national and global water use are described. The need to recycle wastes and waste water is discussed. Problems with present social attitudes toward waste disposal are described. The long term need for new ways to deal with politically sensitive waste disposal issues is discussed. The prevalent NIMBY (not in my back yard) social attitude is described.

Objectives
After studying this section a student should be able to:
    tell why water recycling is actually an old practice
    give reasons why household hazardous waste disposal will continue to be a problem
    explain why conservation and waste reduction will be increasing important in the future
    tell what problems result from the NIMBY response
    explain why water recycling will be more common in the future

Key terms
heavy metals	pesticides	Clean Water Act of 1977
chlorinated organic	monitoring	pollution-discharge regulations
waste reduction	household wastes	politics of water protection
water conservation	protection	water recycling

Suggested Additional Readings

Boyle, Robert H. "The Killing Fields" (toxic drainwater) <u>Sports Illustrated</u> Mar. 22, 1993: 62.

Canby, Thomas Y. "Water Our Most Precious Resource" <u>National Geographic</u> Aug. 1980: 144.

Cantor, Kenneth P. "Water Chlorination, Mutagenicity, and Cancer Epidemiology." <u>The American Journal of Public Health</u> Aug. 1994: 1211.

Colson, Steven D. and Thom H. Dunning Jr. "The Structure of Nature's Solvent: Water" <u>Science</u> July 1, 1994 : p 43.

Cooper, Mary H. "Global Water Shortages: Will the Earth Run Out of Freshwater?" <u>CQ Researcher</u> Dec. 15, 1995: 1113.

Eliassen, Rolf. "Stream Pollution" <u>Scientific American</u> Mar. 1952: 17.

Farvolden, Robert. "Water Crisis: Inevitable or Preventable?" <u>Geotimes</u> July 1995: 4.

"Fit to Drink" <u>Consumer Report</u> Jan. 8, 1990: 27.

Garber, Charri Lou. "Wastewater" <u>ChemMatters</u> April 1992: 12.

Karp, Jopnathan. "Water, Water, Everywhere" <u>Far Eastern Economic Review</u> June 1, 1995: 54.

Knight, Charles and Nancy Knight. "Snow Crystals" <u>Scientific American</u> Jan. 1973: 100.

LaBastille, A. "Acid Rain—How Great a Menace?" <u>National Geographic</u> Nov. 1981: 652.

Mohnen, Volker A. "The Challenge Of Acid Rain" <u>Scientific American</u> Aug. 1988: 30.

Penman, H.L. "The Water Cycle" <u>Scientific American</u> Mar. 1970: 98.

Ward, Fred. "South Florida Water: Paying the Price" <u>National Geographic</u> July 1990: 89.

Answers to Odd Numbered Questions for Review and Thought

1. a. Surface water is water available in rivers, lakes and streams.
    b. Groundwater is water beneath the earth's surface.
    c. A aquifer is a layer of water-bearing porous rock.
    d. Brackish water is fresh water contaminated by salty sea water.
    e. Pollution is defined as any condition that causes the natural usefulness of water, air, or soil to be diminished.

3. a. BOD is the amount of O_2 needed to oxidize organic material in water to CO_2 and H_2O.
    b. Distillation is the process of vaporization of liquid water followed by condensation.
    c. Aeration is the process of mixing air with water to oxidize dissolved or suspended matter.
    d. Sedimentation is the process of separating suspended solids from water by allowing the suspension to stand so solids settle out.

5. Most of the rain water that falls on the United States each day returns to the atmosphere by evaporation or transpiration from plants.

7. Groundwater can be contaminated with pollutants when rainwater runs over or filters through materials, dissolves the pollutants, then percolates into the ground water.

9. Agriculture is the largest single user of water in the United States.

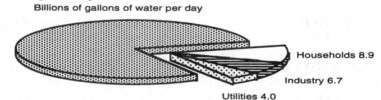

Billions of gallons of water per day

Agriculture 76

Households 8.9

Industry 6.7

Utilities 4.0

11. A dual water system has a system for potable water and another for water for other uses.
     Advantages: consumes less potable water, nonpotable water is less expensive, creates smaller volume of water to be circulated in drinking water system, scale of drinking water system will be smaller and therefore easier to maintain.
     Disadvantages: added cost for second distribution system, dual plumbing needed in homes, potential for cross contamination of the two kinds.

13. Clean water is rarely pure H_2O but is usually a mixture. Dissolved substances come from the atmosphere and the materials the water contacted after it fell to the ground. Substances from the atmosphere are CO_2, N_2, NO_2, O_2, and dust particles. Substances from contact with the ground can be any or all of the following: HCO_3^{1-}, Ca^{2+}, Mg^{2+}, K^{1+}, Na^{1+}, Cl^{1-}, SO_4^{2-}.

15. | Pollutant | Source |
     |---|---|
     | pathogens: | sewage |
     | organic chemicals: | landfills |
     | acids: | mining |
     | radioisotopes: | mine tailings |
     | heat: | cooling towers for electric generating plants |

316

17. Three positive ions often found in natural waters:
magnesium Mg^{2+}, calcium Ca^{2+}, iron(II) Fe^{2+}.
Three negative ions often found in natural waters:
chloride, Cl^{1-}, bicarbonate, HCO_3^{1-} sulfate, SO_4^{2-}.

19. Originally the "Superfund" was a $1.6 billion program for cleaning up hazardous waste sites. The fund cleans up old landfill sites, dump sites, and spill sites so it reduces possible contamination of groundwater.

21. A "manifest" is a document listing hazardous waste by name, amount present, and procedure for its disposal. It identifies the disposer of the material, assigns responsibility, and specifies disposal method.

23. Proposition 65 is a California law that prohibits businesses from knowingly discharging or releasing 200 different hazardous substances into water that may become part of drinking water. Proposition 65 and the Delaney Clause are both "zero risk" concepts.

25. Pure water has a BOD of essentially zero. A high BOD indicates a high concentration of organic matter in the water. This organic matter consumes dissolved oxygen when it undergoes oxidation.

27. The BOD of waste water can be lowered if it is treated with oxygen or ozone. Aeration will mix oxygen into the water and help replace lost oxygen.

29. The gas chromatograph-mass spectrometer is used to measure levels of trace organic compounds in water.

31. Aeration of water reduces levels of dissolved volatile organics because the contact with air allows these organics to vaporize and escape into the atmosphere. Additionally, these organics are oxidized more readily to smaller less hazardous substances when they have greater contact with oxygen in aeration.

33. Biodegradable substances can be easily broken down into simpler smaller molecules by microorganisms while nonbiodegradable substances cannot be easily broken down. A biodegradable detergent is preferred because it should have fewer harmful effects on the environment.

35. DDT, is a persistent polychlorinated hydrocarbon. It is a nonbiodegradable pesticide. It is implicated in interfering with bird reproduction by causing fragile eggs.

37. Chlorine gas, $Cl_2(g)$, is a good oxidizing agent. It kills pathogens by oxidizing structures in the bacterial cell.

39. Bioremediation is the microbial detoxification or degradation of wastes into nonhazardous products. Secondary treatment uses aerobic bacteria to consume organic molecules. Bioremediation and secondary treatment are similar in that both processes consume organic molecules through action of microorganisms.

41. Ammonia, $NH_3(aq)$, and ammonium ion, $NH_4^+(aq)$, can be removed from wastewater by using denitrifying bacteria. These bacteria convert the ammonia and ammonium ion to nitrogen, $N_2(g)$. The unbalanced reaction is: $NH_3(aq)$ or $NH_4^+(aq) \xrightarrow{\text{denitrifying bacteria}} N_2(g)$

43. The soda lime process is used to treat hard water. Sodium ions replace the metal ions Ca^{2+}, Mg^{2+}, Fe^{3+} and Mn^{2+} that make water "hard". The process uses lime ($Ca(OH)_2$) and baking soda ($NaHCO_3$). Calcium is removed because it precipitates as $CaCO_3$. Magnesium is removed because it precipitates as $Mg(OH)_2$. The sodium ion concentration goes up. Water treated with the soda lime process can pose a problem for people who are on a low sodium diet.

45. Chlorination of water containing organic residues can produce chlorinated hydrocarbons. These products can include carcinogenic compounds like dichloromethane, chloroform, trichloroethylene, and chlorobenzene. These compounds are hazardous themselves.

47. Sodium chloride (NaCl) is the most abundant dissolved solid in sea water.

49. Water from salt water is evaporated by solar energy and condensed as pure water. The method has a major disadvantage in that large surface areas are needed for the stills.

51. The "NIMBY" philosophy shifts problems from one area to another and avoids the issue of responsibility. It fails to offer ways to prevent or remedy problems. It essentially denies existence of any problem.

53. Answer depends on individual interpretation.

Speaking of Chemistry

Name _____

Water--Plenty of It, But What Quality?

Across

1 A metal in hard water that can be removed by aeration.

5 Process of mixing of air with water to increase dissolved O_2 and promote oxidation of dissolved matter.

8 One of the two Group IIA metal ions in hard water.

12 Process of microbial detoxification of wastes.

15 A membrane that allows passage of water molecules but not ions or other molecules.

16 Water suitable for drinking.

17 A measure of the amount of dissolved organic matter.

18 Wastewater treatment that removes toxic metal ions, nitrates, ammonium ions, etc.

Down

2 The flow of water from a more dilute solution through a semipermeable membrane into a more concentrated solution.

3 The most abundant substance on earth. It makes up 70% of human body.

4 Any condition that causes the natural usefulness of water, air or soil to be diminished.

6 Abbreviation for the Environmental Protection Agency.

7 An ionic compound like NaCl.

8 The concentration of DDT and hard pesticides is greater in birds and fish because they are at the top of the food _____.

9 A membrane that allows passage of molecules.

10 A hard pesticide first used widely in 1939.

11 DDT accumulates in _____ tissue.

13 Filtration through _____ black is used to remove soluble organic compounds.

14 An oxidizing gas with formula, O_3.

Bridging the Gap   I
Internet access to the Environmental Protection Agency, EPA
Non-point Pollution; Dos and Don'ts Around the Home

The importance of education in bringing nonpoint-source pollution under control is extremely important. The reason for this is pragmatic: What you don't know can hurt the environment. When rain falls or snow melts, the seemingly negligible amounts of chemicals and other pollutants around your home and premises get picked up and carried via storm drains to surface waters. The ramifications include polluted drinking water, beach closings, and endangered wildlife. So what can you do to help protect surface and ground waters from so-called nonpoint-source pollution? You can start at home. Begin by taking a close look at practices around your house that might be contributing to polluted runoff; you may need to make some changes. You can become part of the solution rather than part of the problem of nonpoint-source pollution.

The EPA home page
The EPA home page has the following universal resource locator (URL) address,

> http://www.epa.gov/

The content of any displayed web page can be down loaded to disk as text or it can be printed. The underlined subject headings on the EPA home page are interactive and lead to additional pages. The interactive topics can be selected by clicking on the underlined text. Each one of these active lines leads to another page with a different URL address. The Citizen Information page can be reached if you click on the line."•EPA Citizen Information". This page can be reached directly by using the following URL.

> http://www.epa.gov/epahome/Citizen.html

The "EPA Citizen Information" page
On June 5, 1996 the EPA Citizen Information page looked like the illustration shown on the following page. Remember these pages are constantly updated. The appearance and content of a site will change when the site is updated.

Again this web page has underlined interactive subject headings. Your responsibility is to open the topic " •Dos and Don'ts Around the Home". This page has the following direct URL.

> http://www.epa.gov/OWOW/NPS/dosdont.html

This site contains suggested ways to minimize water consumption and nonpoint-source pollution. The Concept Report Sheet has some questions about the " •Dos and Don'ts Around the Home" information. You are supposed to find the answers and record them on the Concept Report Sheet.

You should also record your opinion of the EPA web page. Please explore the EPA pages and record the name of a page that was interesting to you. Record the URL for this page and tell what you liked about it.

EPA Citizen Information

- •National Drinking Water Week is May 5-11!
- •Water Drops: A Collection Just for Kids
- •Safe Drinking Water Hotline
- •EPA Wetlands Information Hotline
- •WATERSHED '96: Moving Ahead Together
- •You and Clean Water
- •Dos and Don'ts Around the Home
- •Consumer Handbook for Reducing Solid Waste
- •Environment Finance Program Publications
    - •Alternate Financing Mechanisms Report
    - •Funding State Environmental Program Administratio: The use of Fee Based Programs
- •UV Index Document
- •Putting Customers First: EPA's Customer Service Plan
- •Guide to environmental Issues
- •Energy Star Computers Product Listing
- •Volunteer Monitoring Program
- •Protect Your Family From Lead in Your Home
- •Terms of Environment
- •Publications on Radon
- •Superfund Program

[EPA Home Page | Comments | Search | Index]

Bridging the Gap   I
Concept Report Sheet

Name _____

Internet access to the
Environmental Protection Agency, EPA
Dos and Don'ts Around the Home

1. Give the universal resource locator (URL) for the EPA site
    " •<u>Dos and Don'ts Around the Home</u>"

    http://_____

    Name two suggestions the EPA makes that you could adopt.

2. The EPA gives an estimate of the volume of drinking water that one quart of motor can
    pollute. What is this EPA estimate?

3. Give the universal resource locator (URL) for an EPA site you found interesting.

    http://_____

    What makes it interesting to you? Would you recommend it to someone else? You can
    send your opinion to the EPA. They want comments.

Bridging the Gap   II   Name_____

Too Much Water
A Case Study "Waterworld"

"Waterworld" was released in 1995. It stars Kevin Costner, David Finnegan, Jeanne Tripplehorn, and Tina Majorino. The story is set in the future after the earth suffered great climate changes. The polar icecaps have melted and world is covered with water. Civilization as we know it has disappeared. Surviving humans are roaming the seas struggling to find fresh water and food. Conflicting social groups fight over resources and fresh water.

Your assignment has two parts. One part is to view some of or all of "Waterworld" and develop arguments related to the limits of government action dealing with "global warming" and "water" issues. One argument should support the present policy. The other should be an argument in favor of a more intrusive policy. Write your arguments on the Concept Report Sheet.

The second part is to write your opinion about the conditions depicted in the film. Do you think it could ever happen? Why? Why not?

It is probably a good idea to view the movie with a study group. Viewing the movie with a others is probably more fun and it spreads the cost of the rental.

Specifics about the movie.
Produced by: Silver Pictures / Warner Brothers
Released in USA in 1995
Color by: Technicolor
Rating "Waterworld " has a USA "**R**" rating

Bridging the Gap   II   Name _____
Concept Report Sheet
Too Much Water
A Case Study "Waterworld"

1.   Argument in favor of the present policy.

    Argument in favor of a more active water conservation policy.

2.   What is your opinion about the conditions depicted in the film? Do you think it could ever happen? What needs to be done to prevent it?

Solution to Speaking of Chemistry

Water--Plenty of It, But What Quality?

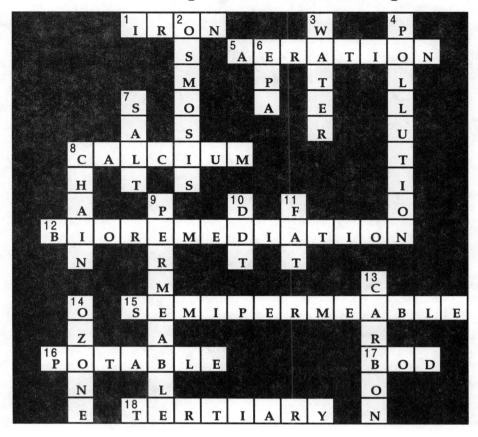

327

328

21 Air--The Precious Canopy

21.1 Doing Something About Polluted Air

The composition and structure of the earth's atmosphere are identified. A brief description of air quality problems is given. The evolution of the Clean Air Act is summarized.

Objectives

After studying this section a student should be able to:

name the level of the atmosphere where weather takes place
name the level of the atmosphere where the protective ozone layer is located
name three naturally occurring atmospheric pollutants
state the purpose of the Clean Air Act of 1970
tell what the Clean Air Act of 1977 did to clean air standards
describe the changes in regulations created by the Clean Air Act of 1990

Key terms

atmosphere	troposphere	stratosphere
mercury vapor	hydrogen chloride	hydrogen sulfide
carbon dioxide	reactive	chlorinated organic compounds
atmospheric pollutants	smoke	carbon monoxide
sulfur dioxide	oxides of nitrogen	1970 Clean Air Act
particulates	ozone	1977 Clean Air Act
sulfur	hydrocarbons	1990 Clean Air Act
volatile toxic substances	stratospheric ozone-depleting chemicals	

21.2 Air Pollutants--Particle Size Makes a Difference

Aerosols and particulates are defined. The types of matter that can be found in particulates are listed. The size of aerosol particles is given in nanometers ($nm = 10^{-9}$ m). Adsorption and absorption are illustrated. Reactions and processes that can occur in an aerosol particle are illustrated. Carcinogenic and mutagenic health hazards posed by aerosols are stated. The effect of particulates on atmospheric temperature is described. Methods to remove particulates from the atmosphere such as scrubbing, electrostatic precipitation, etc., are illustrated and discussed.

Objectives

After studying this section a student should be able to:

give definitions for particulates and aerosols
state the range of diameters for aerosol particles
explain the difference between absorption and adsorption
describe how aerosol particles can pick up hazardous substances and concentrate them
explain how aerosol particles can alter the temperature of the Earth
describe the connection between particulates and volcanic eruptions
tell how particulates and aerosols are naturally removed from the atmosphere
describe how electrostatic precipitators work
tell how filters and scrubbers work to remove particulates and aerosol particles
sketch diagrams showing how adsorption and absorption differ

329

Key terms

particulates

aerosols

nanometer

mutagenic compounds

carcinogenic compounds

adsorb

soot

suspended particulate

gravitational settling

rain and snow

filtration

centrifugal separation

scrubbing

electrostatic precipitation

physical methods

21.3   Smog

A brief history of smog is given. The role of thermal inversions in smog production is discussed. Chemically reducing smog is described. The health effects of chemically reducing smog are summarized. The origins of photochemical smog are described. Primary pollutants, secondary pollutants, free radicals, and the sources of atmospheric hydrocarbons are discussed.

Objectives

After studying this section a student should be able to:

give the definitions for smog, inversion layer, photochemical smog, industrial smog

name the two general types of smog

name the sources of reducing smog

describe the effects of reducing smog

state the definition of a free radical and give an example

give definitions for primary and secondary pollutants

write the equation for the reaction between NO_2 and uv light to form O atoms and NO

write the formula and draw the structures for hydrogen peroxide

write the general formula and draw the general structure for organic peroxides

Key terms

smog

thermal inversion

chemically reducing type

industrial smog

sulfur dioxide

sulfuric acid

asthma

emphysema

respiratory diseases

photochemical smog

oxygenated

chemically oxidizing type

ozonated hydrocarbons

unreacted hydrocarbons

organic peroxide compounds

free radical

primary pollutants

secondary pollutants

21.4   Nitrogen Oxides

The many oxides of nitrogen are described and the general formula NO_x is given. The natural sources of nitrogen oxides are named. Some reactions of NO and NO_2 are illustrated. Photodissociation is defined, and the wavelengths of effective uv light are given. The reaction between water and NO_2 is given and related to aerosol droplet stability.

Objectives

After studying this section a student should be able to:

tell how NO is formed naturally and write the equation for the reaction

identify a natural source of N_2O

describe how NO behaves in the atmosphere

describe the physiological effects of NO_2, nitrogen dioxide

state the definition for photodissociation

tell how ozone is produced from NO_2, uv light, and O_2 (include reactions)

state the wavelength range for uv light that will cause photodissociation of NO_2

write the reaction for the combination of water and nitrogen dioxide and tell what the
    products do to the stability of aerosol drops

Key terms

oxides of nitrogen, NO_x	dinitrogen oxide, N_2O	nitric oxide, NO
nitrogen dioxide, NO_2	parts per billion, ppb	haze
combustion temperature	bronchioconstriction	photodissociation

21.5   Ozone and Its Role in Air Pollution

The formula for ozone is given and its sources are identified. The difference between "good" and "bad" ozone is explained. Ozone as a pollutant in the troposphere is discussed. The 1990 EPA standard for pollutant ozone is stated as 0.12 ppm and discussed in terms of FEV_1 values. The relationship between O_3 and NO_x is explained.

Objectives

After studying this section a student should be able to:

write the formula for ozone

tell the difference between "good" and "bad" ozone

tell why the EPA pollutant level for ozone may be too low

give the definition for forced air volume, FEV_1

describe what ozone does to the FEV_1 of children

explain why ozone concentrations and NO_x concentrations are related

Key terms

ozone, O_3	"good" ozone	"bad" ozone
stratosphere	ultraviolet radiation	forced expiratory volume, FEV_1
NO_x emissions		

21.6   Hydrocarbons and Air Pollution

Natural sources of hydrocarbons like coniferous trees, plants, etc. are identified. Human sources of hydrocarbons such as polynuclear aromatic hydrocarbons (PAH) and the carcinogenic (BAP) are discussed. Efforts to control human sources of hydrocarbon emissions are described. The effect of catalytic converters in automobiles on automobile hydrocarbon emissions is discussed.

Objectives

After studying this section a student should be able to:

name natural sources of hydrocarbon emissions

tell what percent of atmospheric hydrocarbons come from natural sources

explain why hydrocarbons like PAH and BAP are of special concern

describe how automobile hydrocarbon emission rates have changed since 1960

explain why hydroxyl radicals are a pollution problem

Key terms

hydrocarbons	methane gas, CH_4	alkanes
alkenes	alkynes	catalytic converters
PAH	BAP	benzo(α)pyrene
carcinogen	hydroxyl radicals	photodecomposition
polynuclear aromatic hydrocarbons		

21.7 Sulfur Dioxide--A Major Primary Pollutant

The natural and human sources of sulfur dioxide are identified. The effect of SO_2 on the Earth's temperature is explained. The connection between sulfur bearing fossil fuels and atmospheric SO_2 is discussed. Two methods for the removal of SO_2 from exhaust gases are illustrated. The reduction of ground level SO_2 concentrations by diluting exhaust gases using tall stacks is discussed. The reactions of SO_2 to form SO_3 and sulfuric acid are described.

Objectives

After studying this section a student should be able to:

give the formula for sulfur dioxide

explain how SO_2 in the atmosphere can alter the Earth's temperature

identify the human sources of SO_2

describe the EPA requirements proposed for the year 2000 for power plant SO_2 emissions

tell how SO_2 is removed from power plant exhaust gases

explain why power plants that burn sulfur bearing fossil fuels have tall stacks

write the equation for the reaction of $SO_3(g)$ and water to form $H_2SO_4(aq)$

Key terms

sulfur dioxide, SO_2	aerosol droplets	scattering of sunlight
pyrite, FeS_2	calcium sulfite	hydrogen sulfide, H_2S
high-sulfur coal	electrostatic precipitator	tall stacks
500 feet	1000 feet	

21.8 Acid Rain

Acid rain or acid deposition is defined in terms of rainfall pH. The acidic gases SO_2 and NO_x are identified as contributors to acid rain. The source locations of acidic oxide gases and the sites that receive the fallout are mapped for the Northeastern United States. The pH values for regions in the Northeastern United States are illustrated. The maps show the international nature of the acid rain problem. Acid rain is described as responsible for producing "dead" , fishless ponds and lakes. The pH levels that species of fish can tolerate are given. The damage acid rain does to forests and individual trees is described. The effects of acid rain and other atmospheric pollutants on stone and metal structures are discussed. Specific examples of damage to the Tower of London and other historic structures are described. Efforts to prevent and to correct the damage done by acid rain are summarized. The international aspect of acid rain is described as a barrier to solving the problem.

Objectives

After studying this section a student should be able to:

tell why "pure" rain water is acidic and give the normal pH for rainwater

give the pH for neutral water

give the definition for acid rain

explain how gases like CO_2, O_2, and NO_x generate acid rain

describe the problems acid rain creates for fish in ponds and small lakes

tell how acid rain damages trees and forests

describe how acid rain damages stone and metal structures

explain why acid rain is a national and an international problem

explain what the term "fallout" refers to when discussing acid rain

Key terms

acid rain	pH below 5.6	nitric acid, HNO_3
nitrous acid, HNO_2	sulfuric acid, H_2SO_4	sulfurous acid, H_2SO_3
"dead" ponds and lakes	dying or dead forests	acid rain tree damage

21.9   Carbon Monoxide

The natural and human sources of carbon monoxide are described. Natural sources of CO are described as ten times greater than human sources. The major source of human generated carbon monoxide is identified as automobile exhaust. The level of atmospheric CO is described as constant and not really explainable.

Objectives

After studying this section a student should be able to:

name the natural sources of CO
name the biggest artificial source of CO
tell how the amounts of natural and human generated CO compare
write the equation for the reaction between insufficient oxygen and coal to yield CO
explain why automobile catalytic converters play a role in limiting CO emissions

Key terms

carbon monoxide	natural sources	oxidation of decaying organic matter
industrial sources	automotive sources	gasoline engines
catalytic converters		

21.10   Chlorofluorocarbons and the Ozone Layer

Chlorofluorocarbons (CFCs) are defined and examples are illustrated. Uses for CFCs are described along with a brief history of their development. The photochemical reaction of CFC-11 is used as an example of how CFCs can be a source of Cl· and ClO· radicals. The role of ozone as a uv absorber is described. Ozone depletors are identified and the steps in the chain reaction in ozone depletion are discussed. The ozone hole is described, and the consequences of increased ozone intensity at sealevel are explained. Natural ozone depletors are discussed. Efforts to control CFC emissions are summarized. The Montreal Protocol is described. Alternates to CFCs are discussed. The reasons and economic costs of banning CFCs are discussed.

Objectives

After studying this section a student should be able to:

write the formula for ozone
give the definition for CFCs and an example
explain why CFCs were used as refrigerants, degreasers, etc.
describe how photodissociation makes CFCs a threat to stratospheric ozone
write the reaction for the photodissociation of CFC-11
tell why stratospheric ozone is "good"
explain what an "ozone hole" means
describe the role of ozone in absorbing ultraviolet light
describe the health problems that will result from ozone depletion
summarize the chronology of the political actions to restrict CFCs
describe compounds such as perfluorobutane that are being used in place of CFCs
tell why naturally occurring CH_3I is able to be an ozone depletor
describe the likely future for CFC usage

Key terms

CFCs	chlorofluorocarbons	halogenated hydrocarbons
refrigerant gas	degreasers	propellant
CFC-11	CFC-12	ozone-depleting substances
stratospheric ozone	ultraviolet light	black-market
200-300 nanometer range	photodissociation	chlorine oxide, ClO
Montreal Protocol	ozone hole	hydrochlorofluorocarbons
retrofitted	HCFC-134a	HCFC-141b
halons	bromodifluoromethane	bromine oxide radicals, BrO
perfluorobutane	"per-"	chlorotetrafluoroethane

21.11 Carbon Dioxide and the Greenhouse Effect

The trends in atmospheric CO_2 concentrations are graphed for the period 1958 to 1990 and discussed. The link between increased fossil fuel use and CO_2 concentrations is described. The energy balance between solar energy received by the earth and energy radiated into space is described. The changes in CO_2 concentrations are related to the greenhouse effect. The greenhouse effect is illustrated and the process is explained. Greenhouse gases like O_3, CH_4, H_2O, and CO_2 are identified. Projected increases in atmospheric temperature are linked to CO_2 concentrations above 360 ppm. Political difficulties in solving the global warming problem are discussed. International efforts to control CO_2 emissions to minimize global warming are described.

Objectives

After studying this section a student should be able to:

describe the trend in atmospheric CO_2 levels since 1958
explain why CO_2 levels have increased
tell how CO_2 concentrations influence atmospheric temperature
give the definition for a greenhouse gas
explain what the greenhouse effect is
name four greenhouse gases
give the definition for global warming
tell how greenhouse gases alter the atmospheric temperature
explain why controlling the greenhouse effect is an international problem
describe the political problems associated with controlling global warming
summarize the United Nations projections for greenhouse gas concentrations

Key terms

fossil-fuel	carbon dioxide, CO_2	global concentration of CO_2
cut-and-burn	OPEC	solar radiation
visible light	reradiate	infrared
water vapor	ozone	methane, CH_4
greenhouse effect	greenhouse gases	atmospheric temperature
absorbing blanket	global warming	global industrial emissions

21.12 Industrial Air Pollution

Industrial chemical releases are described and compared to natural chemical releases. The need for industrial disclosure of chemical releases is discussed. The five states with the greatest amount of chemicals released are identified. The United States EPA requirement for industrial release reports and Community-Right-to-Know regulations are discussed. The EPA hotline for information about reported chemical releases is given. 1-800-535-0202.

Objectives
After studying this section a student should be able to:
> describe the effects of the chemical release that occurred in Bhopal, India
> tell how the Bhopal incident influenced the EPA release reports
> give the approximate number of compounds that industry must file release reports for
> explain the reasoning behind the Community-Right-to-Know regulations
> give the EPA toll free number that gives information about reported chemical releases

Key terms

industrial releases	summary of annual releases	Superfund
release report	Community Right To Know	solvents
NO_x	SO_2	CO_2
CFCs	metal particulates	acid vapors
unreacted monomers	EPA	1-800-535-0202

21.13   Indoor Air Pollution
The EPA studies of indoor air are described. The sources of indoor pollutants are identified. The move to make homes energy efficient and air tight is identified as one reason for indoor pollution. A schematic of a typical home is used to show the types and sources of indoor pollutants. Indoor pollution in nonsmoker homes is compared with pollution in the homes of smokers. The "sick building syndrome" is described.

Objectives
After studying this section a student should be able to:
> give reasons why indoor pollution is more of a problem than in previous years
> name typical indoor pollutants and identify their sources
> tell how indoor pollutants differ in a nonsmoker's home from a smoker's home
> explain what is meant by the term "sick building syndrome"

Key terms

indoor air pollution	energy-efficient homes	carcinogen
benzene	tobacco smoke	radon-222
gasoline	formaldehyde	*para*-dichlorobenzene
perchloroethylene	carbon monoxide	methylene chloride
asbestos	nitrogen oxides	chloroform
fungi	bacteria	sick building syndrome
building materials	consumer products	outside air

Additional readings
Beardsley, Tim. "Add Ozone to the Global Warming Equation" <u>Scientific American</u> Mar. 1992: 29.

Calvert, Jack G., et al. "Achieving Acceptable Air Quality: Some Reflections on Controlling Vehicle Emissions" <u>Science</u> July 2, 1993: 45.

Greene, David L. and K. G. Duleep. "Costs and Benefits of Automotive Fuel Economy Improvement: A Partial Analysis" <u>Transportation Research</u> May 1993: 217.

Matthews, Samuel W., and James A. Sugar. "Under the Sun: Is Our World Warming?" <u>National Geographic</u> Oct. 1990: 66.

335

Mohnen, Volker A. "The Challenge of Acid Rain" <u>Scientific American</u> Aug. 1988: 30.

Viggiano, A.A., et al. "Ozone Destruction by Chlorine; the Impracticality of Mitigation Through Ion Chemistry" <u>Science</u> Jan. 6, 1995: 82.

Washington, Warren M. "Where's the Heat?" <u>Natural History</u> Mar. 1990: 67.

Zimmer, Carl. "Unintended Consequences" <u>Discover</u> Mar. 1995: 32.

Answers to Odd Numbered Questions for Review and Thought

1.   a. Polluted air is air that contains unwanted and harmful substances.
     b. Particulates are solid particles in the air with diameters greater than 10,000 nm.
     c. Aerosols are mixtures of water droplets and particulates with diameters in the range of 1 nm - 10,000 nm.
     d. A thermal inversion is a layering of air with cold high, density air near the earth's surface and warmer, low density air layered above.

3.   a. Acid rain is rainwater with a pH lower than 5.6.
     b. CFC is the abbreviation for chlorofluorocarbons such as CFC-11, CCl_3F.
     c. The ozone hole is a region the stratosphere that has lower than normal ozone concentration. This lowered concentration level results in a more UV light reaching the earth's surface.
     d. A greenhouse gas is a molecule that absorbs infrared light and radiates it to the atmosphere.
     e. Global warming refers to a worldwide increase in atmospheric temperature.

5.   Clean Air Act has the abbreviation CAA. The first act was passed in 1970 and originally controlled air pollution from cars and industry. Amendments in 1977 imposed stricter auto emission standards. The 1990 CAA extends to manufacturing and commercial activity. This most recent version of the CAA regulates particulates, ozone, carbon monoxide, oxides of nitrogen and sulfur, carbon dioxide, and substances that would deplete stratospheric ozone.

7.   Industrial smog is different from photochemical smog because it is chemically reducing while photochemical smog is chemically oxidizing. Industrial smog contains sulfur dioxide mixed with soot, fly ash, smoke, and partially oxidized organic compounds. Photochemical smog is essentially free of SO_2, but contains ozone, ozonated hydrocarbons, organic peroxide compounds, nitrogen oxides, and unreacted hydrocarbons.

9.   Ozone is formed in the stratosphere when uv light breaks up O_2 to produce oxygen atoms; these react with additional O_2 to form O_3, ozone.

11.   Of the of nitrogen oxides "NO_x" in the atmosphere, 97% come from natural sources. Lightning strikes during electrical storms produce NO. Some bacteria produce N_2O. Nitric oxide, NO, is so reactive that it combines with O_2 in the atmosphere to produce NO_2. About 3% of the atmospheric nitrogen oxides come from human activity such as combustion in automobile engines.

13. Volcanic eruptions can contribute to global cooling because the eruption will throw dust particles into the air. These dust particles scatter and reflect sunlight into space, so the solar energy never reaches the earth's surface. This will decrease the amount of energy striking the earth and decrease the atmospheric temperature.

15. Hydrocarbons are released into the atmosphere by natural sources and human sources. Hydrocarbons are put into the atmosphere by living plants such as deciduous trees, by the decay of dead plants and animals, and by excrement from insects and animals. Human activity introduces hydrocarbons when organic solvents are used, for example in the handling of petroleum products, etc. Generally only human activities are within our control.

17. Nitrogen dioxide plays a role in the formation of ozone in the troposphere.

19. Nitrogen dioxide (NO_2) will dissociate to form an oxygen atom and nitric oxide, if it is excited by a sufficiently energetic photon of uv light. The O atom forms O_3 by reaction with O_2 . $NO_2(g) + hv ---> NO(g) + O(g)$
$$O + O_2 ---> O_3$$

21. All three are oxidizing agents. Ozone, sulfur dioxide and nitrogen dioxide cause lung damage.

23. Ozone is produced following decomposition of NO_2 and subsequent reaction of the O atom with O_2 to form O_3, so controlling the NO_x emissions will control ozone. (see question 19 above)

25. Automobile catalytic converters control carbon monoxide (CO) and nitrogen oxides(NO_x).

27.

1960 (no catalytic converters)	g/mile	1993 (catalytic converters)	g/mile	1995 (estimated)	g/mile
HC	10.6	HC	0.41	HC	0.41
CO	84.0	CO	3.4	CO	3.4
NO_x	4.1	NO_x	1.0	NO_x	1.0

All three categories of emissions decreased on a per mile basis because of catalytic converters. Lead emissions decreased because unleaded gasolines are now used. These data show what happens per mile of automobile use. What these do not show is how the totals compare because of changes in the number of miles driven. The cars are cleaner, but increases in total miles driven could increase actual pollution.

29. Oil burning electric fuel plants generate SO_2. The amount of SO_2 emitted can be reduced by either using low sulfur fuel oil or by passing plant exhaust gas through molten sodium carbonate to form sodium sulfite.

31. Rain tends to be acidic because it dissolves CO_2 to make H_2CO_3. This normally creates a solution with a pH of 5.6. Human sources put SO_2 and NO_x into the atmosphere. They react with rainfall to yield nitric acid, sulfuric and sulfurous acid. The pH of these mixtures will be lower than the natural 5.6. This precipitation is called acid rain. Values have been observed in the pH range of 4.0 to 4.5. Isolated storms have yielded rain with a pH of 1.5.

33. The reaction between carbon and oxygen when there is insufficient oxygen is
$2 C + O_2 \longrightarrow 2 CO$ The product is carbon monoxide.

35. CFCs are linked to depletion of ozone concentrations in the stratosphere. These regions of lowered ozone concentration are relative ozone "holes". These ozone holes lead to increased ultraviolet light levels at sea level resulting in greater rates of skin cancer. Some unforeseen effects on algae, plants and animals that could upset the ecological balance are also causes for worry.

37. Ultraviolet light can cause the break up of oxygen molecules. $O_2 + hv \longrightarrow 2 O$
The oxygen atoms formed by the dissociation of O_2 can form ozone in the following reaction. $O_2 + O \longrightarrow O_3$

39. Automobile air conditioners primarily used CFC-11, CCl_3F. CFCs can circulate in the atmosphere and reach the stratosphere. There ultraviolet light can break the carbon chlorine bond to produce Cl atoms. These Cl atoms can react with O_3 to deplete ozone concentrations in the stratosphere.

41. Methyl iodide (CH_3I) is released into the atmosphere by marine organisms. The chapter does not include any other natural halocarbons. The question should ask for one compound not three.

43. Greenhouse gases absorb infrared radiation. This prevents it from radiating into space. The excited molecules reradiate the absorbed energy into the atmosphere. This results in an increase in the atmospheric temperature.

45. Carbon dioxide is removed from the atmosphere by photosynthesis and dissolving in sea water.

47. Natural processes release large amounts of pollutants but usually are part of an ongoing natural cycle and equilibrium. There is nothing we can do about the natural. In some cases human sources can be eliminated by changing processes. Human sources are smaller, but like the proverbial straw that broke the camel's back, they can have profound effects.

49. Each community is different so there are many possible right answers.

Speaking of Chemistry

Name _____

Air -- The Precious Canopy

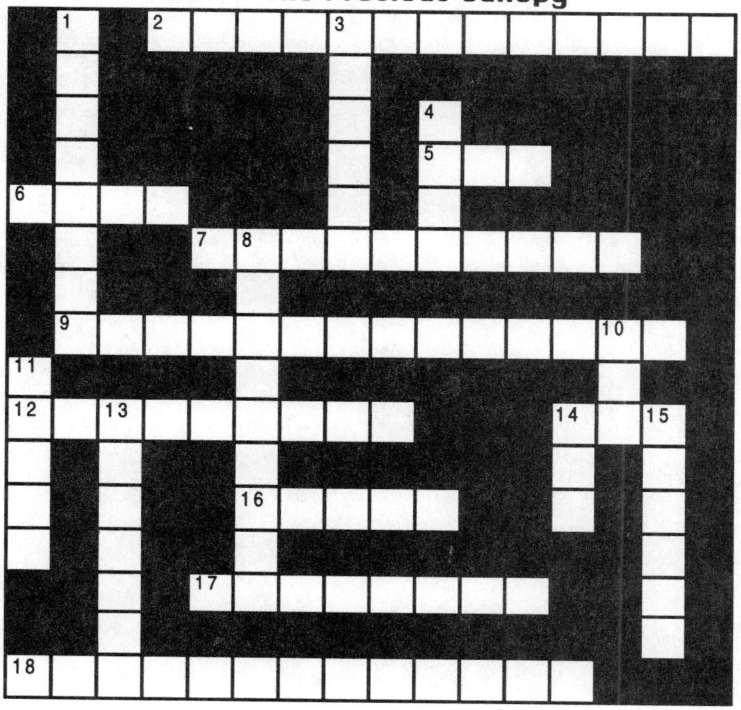

Across

2 Increased atmospheric temperature resulting from greenhouse effect.

5 Forced Expiratory Volume

6 Smoke, dust and _____ are suspended particulates.

7 Type of reactions that produce energy from burning fossil fuels in auto engines.

9 Compounds like NO, ⊙=⊙,nitrogen oxide and nitrogen dioxide, NO_2, ⊙-⊙.

12 Aerosol particles may contain _____ compounds that cause mutations.

14 Polynuclear aromatic hydrocarbons like benzo(α) pyrene.

16 Reactive gas that absorbs ultraviolet light in the stratosphere.

17 Photochemical smog contains ROOR a reactive organic _____ that is related to

hydrogen peroxide, HOOH, ⊙-⊙.

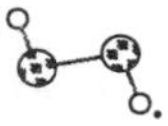

18 Reacts with water to form sulfuric acid and sulfurous acid.

Down

1 The most abundant gas in the atmosphere.

3 Aerosol particles can _____ and concentrate pollutants on their surfaces.

4 Compounds linked to ozone depletion.

8 Caused by decrease in stratospheric ozone.

10 Environmental Protection Agency

11 Fog and _____ are examples of aerosols.

13 _____ inversion: mass of warm air over a mass of cooler air.

14 Parts per billion or 1/ 1,000,000,000

15 Class of compounds used in fire

extinguishers like CF_3Br 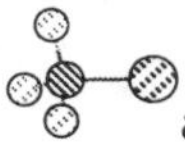 and CF_2BrCl,

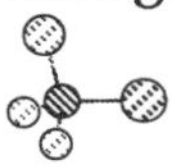

Bridging the Gap   I

Internet access to the United States Geological Survey, USGS
National Water Conditions, pH of Precipitation,
Acid Rain Where You Live

The USGS : National Water Conditions

The United States Geological Survey is a part of the Department of the Interior. Specific activities of the USGS have separate Internet home pages. The monitoring of the National Water Condition is one of the responsibilities of the USGS. The "Table of Contents" page for the various activities is readily accessible using the following universal resource locator (URL) web site address.

http://h2o.usgs.gov/nwc/NWC/html/TOC.html

pH of Precipitation: map of the United States

You can scroll down the table of contents page and click on the active line named

"•pH of Precipitation".

This will give you the "pH of Precipitation" page. You can go to this page directly if you use the following URL

http://h2o.usgs.gov/nwc/NWC/pH/html/ph.html

This page gives a map of the United States with state boundaries. The pH values are updated every month. Data from earlier reports can be also accessed. There are 190 monitoring sites on the map. The point and click feature of the map lets you choose a state and get rainfall pH values from the sites in that state. You can either print the image or save it to disk.

The purpose of this assignment is to open the "pH of Precipitation" page under "National Water Conditions" and select a state of your choice. You are to record the name of the state, sketch its outline (alternately you can download the map). You are to record the pH readings in that state.

You are also to note the pH values for the states east of the Mississippi River. You are to find the state in this group with the lowest value pH for its precipitation and record both the pH and the identity of the state.

You are supposed to record the values for pH of rainfall in Northern Arizona and the pH for points in Ohio. If there are differences, you are supposed to give some explanation for them.

341

Bridging the Gap I
Concept Report Sheet
Internet access to the
United States Geological Survey, USGS

Name _____

1. Give the universal resource locator (URL) for the USGS "pH of Precipitation" page that shows a map of the United States and pH values across the country.

   http://_____

   What state east of the Mississippi has the most acidic rainfall? (the lowest pH)

2. Identify the state you chose to check. What are the pH values for sites in this state? Are the reported values more acid than normal rainfall with a pH of 5.6?

3. Give the universal resource locator (URL) for the state you selected.

   http://_____

4. What is the pH for rainfall in Northern Arizona?

   What is the pH for rainfall in Ohio?

   Which of these is more acidic?

   What environmental reasons exist for the differences?

Bridging the Gap   II   Name _____
Concept Report Sheet
Global Warming and Individual Responsibility

The global warming problem is tied to greenhouse gases. Natural greenhouse gases are beyond human control. The artificial sources of greenhouse gases are controllable, at least in theory.

The social and political problems of reducing greenhouse gas emissions are tremendous. You probably read in the chapter that the global population is projected to be about 11.5 billion people in the year 2050. This projection means a doubling of the global population.

There are studies that say the temperature of the atmosphere might remain constant if CO_2 concentrations in the atmosphere can be kept at 360 ppm. Do you think a doubling of the Earth's population can happen without exceeding this CO_2 concentration? Your assignment is to give an answer to this question and justify your answer.

Your second assignment is to meet with other members of your class and discuss what will happen to CO_2 concentrations if people continue their present patterns of fossil fuel use. Your discussion should be aimed at developing two suggestions for changes in individual behavior that will help keep CO_2 concentrations from rising .

Use this page to record your suggestions and summarize your group's discussion.

Solution to Speaking of Chemistry crossword puzzle

Air -- The Precious Canopy

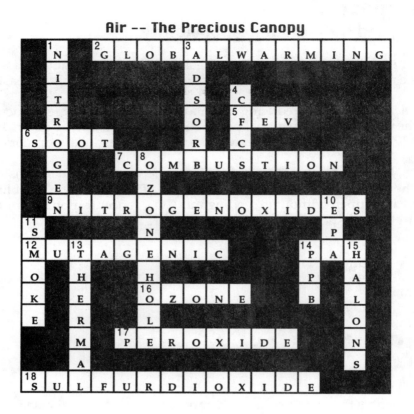

22 Feeding the World

22.1 World Population Growth

International conferences on global population are listed. The current global population is given as approximately 5.9 billion. The rate of global population growth is described. The year 2050 is given as the time when the world population will be double what it is now. Estimates are given for the number of people currently living in poverty.

Objectives
After studying this section a student should be able to:

state the approximate global population in 1995
give number for the projected global population in 2050
give an estimate of the percentage of the global population that lives in poverty
explain why there is concern about global population growth and agricultural productivity

Key terms

Worldwatch Institute   Earth Summit   Cairo, September, 1994
Rio de Janeiro, 1992   world population   human population growth
agrichemicals   risk versus benefit   agricultural productivity
rate of growth of world's population
United Nations Conference on Environment and Development
United Nations International Conference on Population and Development

22.2 What is Soil?

The structure of soil is described, and the four components of soil are listed. The function of humus is described. The structure of soil is illustrated and labeled. Definitions are given for horizons, humus, friable soil, topsoil, subsoil, separates, and loam. The differences between normal air and air mixed in soil are explained. The effect of elevated CO_2 levels in soil on the soil's pH is described. Sweet and sour soils are defined. The percolation process for water movement in soil is described. Leaching is defined; selective leaching of soil is described and its effect on salts and metal ion concentrations is explained.

Objectives
After studying this section a student should be able to:

name the four components of soil
name the sources of organic matter in soil
explain how humus makes a soil more friable
give definitions for humus, horizons, topsoil, subsoil, loam
describe how oxidation of organic matter in soil alters it's pH
describe the difference between sweet soil and sour soil in terms of pH
describe the three ways water is stored in soil
give the definitions for absorbed water and adsorbed water, and tell how they differ
describe the percolation process for water movement through soil
explain how leaching of soil affects concentrations of salts of Group IA and IIA metals
describe how leaching of soil affects the concentrations of salts of Group IIIA
explain what is meant by selective leaching, tell how it affects soil pH
name two processes that influence soil pH
sketch and label a diagram of soil showing horizons

Key terms

soil	mineral particles	organic matter
water	air	humus
soil structure	friable	horizons
topsoil	subsoil	salts
clay	silt	sand
gravel	separation rates	loam
air in soil	normal dry air	groundwater
acidic soil	sour soil	sweet soil
absorbed	adsorbed	pores
transpire	evaporate	salts of Group IA & IIA metals
percolation	waterlogged soil	salts of Group IIIA metals
plant products	selective leaching	salts of transition metals
leaching effects	solubility	alkaline

22.3   Nutrients

Elemental nutrients are classified as nonmineral and mineral. The nonmineral nutrients are identified as carbon, hydrogen and oxygen and can be obtained from the atmosphere. Mineral nutrients are subdivided into primary, secondary and micronutrients. They are available to plants as solutes through the plant roots. Primary nutrients are identified as nitrogen, potassium, and phosphorus. The water soluble forms of primary nutrients are described. The effect of pH on the solubility of phosphate is described. The solubilities of secondary nutrients, calcium and magnesium, are discussed. Iron is defined as a micronutrient and the effect of pH on its solubility is explained.

Objectives

After studying this section a student should be able to:

explain how nonmineral and mineral nutrients differ
name the three nonmineral nutrients
explain why mineral nutrients must be water soluble to be used by a plant
give the definition for nitrogen fixation
describe the role of nitrogenase in nitrogen fixation by legumes
describe how pH influences the solubility of phosphate
describe the symptoms of chlorosis and tell what causes it

Key terms

elemental nutrients	nonmineral nutrients	mineral nutrients
primary nutrients	secondary nutrients	micronutrients
nitrogen fixation	oxidize nitrogen	nitric oxide, NO
nitric acid, HNO_3	nitrogenase	soluble mineral
nitrate salts	$H_2PO_4^-$	dihydrogen phosphate ion
K^+	HPO_4^{2-}	monohydrogen phosphate ion
Ca^{2+} and Mg^{2+}	sulfate ion, SO_4^{2-}	trivalent phosphate ion, PO_4^{3-}
chlorophyll	chlorosis	complex ions

22.4   Fertilizers Supplement Natural Soils

A brief history of fertilizer development is given. The usable life of land from slash-burn-cultivate cycle is estimated to be only one to five years. The costs of applying chemical fertilizers worldwide are discussed. Productivity of chemically fertilized land is compared with unfertilized land. The labeling system for chemical fertilizers is explained. Examples are given for calculations of weights of N, P and K for labeled fertilizers. Quick release and slow release fertilizers are defined. Nitrogen fertilizers like urea, ammonia, and ammonium nitrate are identified. Production and use of superphosphate fertilizer is discussed.

Objectives
After studying this section a student should be able to:

> tell how long farmland used in the slash-burn-cultivate cycle can produce crops
> state definitions for straight, complete, and mixed fertilizers
> tell why ammonia, urea, and ammonium nitrate are applied to soil
> tell what a fertilizer grade label indicates about nutrient concentrations
> name the compound made in the Haber process
> use a fertilizer label to calculate
>> the weight of nitrogen present as N,
>> the weight of phosphorus present as P_2O_5 and
>> the weight of potassium as K_2O
> explain why "superphosphate" is used as a fertilizer instead of $Ca_3(PO_4)_2$

Key terms

slash-burn-cultivate cycle	legumes in crop rotation	mixed fertilizer
straight fertilizer	complete fertilizer	grade
phosphorus as P_2O_5	potassium as K_2O	potash
quick-release fertilizer	slow-release fertilizer	"phosphate"
anhydrous ammonia	Haber process	urea
explosives	"liquid nitrogen"	ammonium nitrate, NH_4NO_3
$Ca_3(PO_4)_2$	phosphate rock	superphosphate

22.5   Protecting Plants in Order to Produce More Food

Pesticides, insecticides, herbicides, and fungicides are discussed. The economic impact of these compounds is explained. A summary of the historical development of insecticides is given. The development of DDT is discussed. Benefits and problems with persistent chlorinated hydrocarbon pesticides are discussed The LD_{50} values for various pesticides are compared. Environmentally friendly pesticides are described. Tillage is defined. Selective and nonselective herbicides are discussed. The enzyme inhibiting herbicides like glyphosate are discussed. The widely used herbicide 2,4-D is described. The reasons for banning 2,4,5-T are given. Pre-emergent herbicides like paraquat are described. Traditional fungicides like copper and sulfur compounds are discussed. New fungicides such as Thiram are described.

2, 4, 5-trichlorophenoxyacetic acid

2, 4-dichlorophenoxyacetic acid

DDT

Thiram

Objectives

After studying this section a student should be able to:

  give the definitions for herbicides, fungicides, insecticides, and pesticides
  explain what is meant by the term "persistent insecticide"
  describe the history of DDT and tell why it is banned in the United States
  explain what makes carbamate insecticides attractive
  describe the reasons why insecticides are used even though they may pose other problems
  explain what it means when a substance is said to have a low LD_{50}
  tell why 2, 4, 5-T is banned by the EPA
  give the definition for tillage and explain why herbicides make it less necessary
  give an example of a common herbicide
  describe how glyphosate functions as a herbicide
  name a pre-emergent herbicide
  explain why fungicides are necessary
  name an "old" fungicide and a "new" fungicide

Key terms

herbicides	biological half-life	World Health Organization
fungicides	arsenic compounds	petroleum oils
insecticides	nicotine	hydrogen cyanide gas
pyrethrum	rotenone	organophosphorus compounds
toxicity	cryolite	anticholinesterase poison
persistent pesticides	malaria	metabolized
biodegradable	DDT	nonselective herbicide
carbamate	Malathion	fat soluble
LD_{50}	pirimicarb	selective insecticide
selective herbicide	tillage	minimum tillage
growth hormones	arsenates	calcium cyanamide, CaNCN
borates	arsenites	sulfates
Agent Orange	chlorates	Dioxin
triazines	Benomyl	dithiocarbamates
glyphosate	Round-up	Benlate
sulfonylureas	Oust	inhibiting plant enzymes
valine	leucine	essential amino acids
Paraquat	pre-emergent herbicide	isoleucine
elemental sulfur	copper compounds	mildew
2,4-D, 2,4-dichlorophenoxyacetic acid		mercury compounds
2,4,5-T, 2,4,5-trichlorophenoxyacetic acid		

22.6 Sustainable Agriculture

The magnitude of the topsoil erosion problem is described. Deficiencies in farming practices are identified. Pest resistance to pesticides and soil compaction are two problems resulting from poor practices. Sustainable and organic farming are discussed as alternatives to poor farm practices. National Research Council recommendations for new farming practices are given. Integrated pest management, IPM, is described as a way to reduce pesticide usage. Natural insecticides, insect pheromones, disease-resistant crops, and insect pest predators are discussed as parts of the IPM program. Natural insecticides like pyrethrum and neem-tree seed oil are described. Natural selective insecticides based on bacterial toxins are discussed.

Objectives
After studying this section a student should be able to:
- describe the scale of soil erosion in Iowa in the past 150 years
- explain what soil compaction does to soil fertility
- describe the main aspects of organic farming
- state the six proposals in the 1989 National Research Council report
- describe features of the integrated pest management system
- explain why natural insecticides are desirable
- describe the benefits of pyrethrum and pyrethrin insecticides
- explain why pyrethrum must be mixed with other insecticides to enhance its effectiveness
- describe the advantages offered by insecticides based on *Bacillus thuringiensis* (*Bt*)
- describe test results for California studies using neem-oil based Margosan-O
- explain why rotating crops helps fight insect infestations

Key terms
organic farming	alternative agriculture	sustainable agriculture
natural fertilizers	green manure	biological pest controls
insect pheromones	natural insecticides	integrated pest management
pyrethrum	bacterial toxins	*Bt* strains
Bt toxins	biopesticides	azadirachtin
neem-tree seed oil		

22.7   Agricultural Genetic Engineering
Transgenic crops are defined. The development time for natural breeding methods is compared with the development time for genetically engineered plants. The ability to engineer plants with specific traits is described. Genetically engineered traits in unlimited numbers are suggested. Tomatoes resistant to tobacco mosaic virus and plants that produce their own insecticides are examples of successful genetically engineered crops.

Objectives
After studying this section a student should be able to:
- give the definition for transgenic plants or crops
- tell why developing genetically engineered crops is quicker than traditional breeding
- tell why people oppose genetic engineering experiments with plants
- give an example of a successful genetically engineered crop

Key terms
transgenic crops	genetically engineered	natural breeding
FlavrSavr tomato		

Additional Readings
Christensen, Damaris. "Microorganisms Used to Fight Fruit Rot" Science News June 4, 1994: 359.

Cooper, Mary H. " Regulating Pesticides: Do Americans Need More Protection From Toxic Chemicals?" CG Researcher Jan. 28, 1994: 75.

Hanson, David. "Pesticides and Food Safety: Administration Proposes Broad Reforms" Chemical and Engineering News Jan. 28, 1994: 6.

Newman, Alan. "Would Less Frequent Pesticide Monitoring be Better?" Environmental Science & Technology Feb. 1995: 70.

"Pest Wars" <u>National Geographic</u> April 1992: 13.

Raloff, Janet. "Berry Scent Defends Fruit From Fungus" <u>Science</u> Aug. 7, 1993: 84.

Safrany, David R. "Nitrogen Fixation" <u>Scientific American</u> Oct. 1974: 64.

Schneider, Dietrich. "The Sex-attractant Receptor Of Moths" <u>Scientific American</u> July 1974: 28.

Sheldon, Richard P. "Phosphate Rock" <u>Scientific American</u> June 1982: 5.

Answers for Odd Numbered Questions for Review and Thought

1. The earth's carrying capacity depends on the amount of productive land, agricultural practices, biotechnology advances, the rate of degradation of farmland global pollution, the amount of water for irrigation, and the level of an acceptable quality of life.

3. Soil is a mixture of mineral particles, organic matter, water, and air.

5. Soil has a structure made up of a series of layers called horizons which are loosely packed and permeable near the surface and gradually change to impermeable solid rock. The upper most horizon or topsoil is a permeable mixture of living organisms and humus. The next horizon down is the subsoil which is permeable or friable clay in the upper layer and stiff clay below. The deepest horizon is the substratum which has an upper level of soft rock and a lower layer of solid rock.

7. Limestone is $CaCO_3$. It is a basic compound and reacts with acids, H^+, to produce bicarbonate ion and Ca^{2+} ion. Limestone will neutralize acids in the soil. Adding a base such as limestone to soil causes the pH to go up. This corresponds to a sweet soil. Low pH values below 7 indicate acidic conditions, 7 indicates a neutral condition and values above 7 indicate basic.

9. The three nonmineral nutrients that plants need for healthy growth are carbon, hydrogen and oxygen. Plants get carbon from carbon dioxide in the atmosphere. They get hydrogen and oxygen from water.

11. a. Metal ions from Group IA and Group IIA are more easily leached.
    b. The pH decreases because the more highly charged metal ions like Ca^{2+}, Mg^{2+} aluminum (Al^{3+}) and the transition metals are strongly attracted to water molecules. The attraction between the Al^{3+} and the oxygen in water weakens the O-H bond. The O-H bond is weakened so much that it sometimes breaks to release H^+ ions into the soil. This is illustrated below. The result is a more acid soil, and the pH goes down.

13. a. Nutrients are substances needed by green plants for healthy growth. There are 18 known elemental nutrients.

b. Nonmineral nutrients are carbon, oxygen and hydrogen. They are available from water and atmospheric carbon dioxide.

c. Mineral elemental nutrients are water soluble substances that plants can only absorb as solutes through their roots.

Nonmineral	Primary	Secondary	Micronutrients
Carbon	Nitrogen	Calcium	Boron
Hydrogen	Phosphorus	Magnesium	Chlorine
Oxygen	Potassium	Sulfur	Copper
			Iron
			Manganese
			Molybdenum
			Vanadium
			Zinc

15. A soil shortage of N, P and K is more likely than a soil shortage of Ca, Mg and S. The nutrients N, P, and K are primary elemental nutrients and are needed in greater amounts. The elements N and K are typically present as very water soluble substances. They are easily leached from the soil, so this is another reason.

17. The balanced equation for the reaction in the Haber process is:
$N_{2(g)} + 3 H_{2(g)} \rightleftharpoons 2 NH_{3(g)}$ This is a method to fix nitrogen from atmospheric nitrogen. It is the synthetic source of ammonia. The process is important because it provides cheap ammonia for agricultural fertilizer production on a huge scale.

19. Legumes are a class of plants like soybeans and alfalfa. These plants have nitrogen fixing bacteria living in their roots. Nitrogenase is an enzyme found in these nitrogen fixing bacteria. It catalyzes reactions that convert nitrogen (N_2) to ammonia (NH_3)under atmospheric conditions.

21. Chlorosis is a plant condition of low chlorophyll. It is caused by a deficiency of any one of the three nutrients: magnesium, nitrogen or iron. This deficiency leads to low chlorophyll content. A symptom of chlorosis is the presence of leaves that are pale yellow instead of green.

23. The fertilizer labeling system indicates the percentage by weight of nitrogen (N), phosphorus (P_2O_5) (phosphate) and potassium as K_2O (potash). None of the pure elements are present in the fertilizer. Each element is present in some compound. None of the reference substances are present in the fertilizer bag either.

25. Urea, H_2NCONH_2, decomposes in the soil to form ammonia.

27. a. A quick-release fertilizer dissolves readily in water. It dissolves easily and can be picked up quickly by the plant roots.
b. A slow-release fertilizer is not as water soluble. It dissolves very slowly in water and it is available to the plant more slowly, so it is taken up slowly by the plant.

29. Ammonium nitrate is inexpensive and a good source of water soluble nitrogen. Farmers benefit because cheap fertilizers keep food production costs down. Consumers benefit because food prices are low. The dark side of ammonium nitrate is that it is an explosive.

Handling it can lead to industrial accidents. It can be used as an explosive by terrorists as was done in 1995 when approximately a ton of ammonium nitrate was detonated next to the Federal Building in Oklahoma City.

31. DDT is a nonpolar organic compound. It is fat soluble because "like dissolves like" and both are nonpolar. DDT poses problems because it is a carcinogen and does not degrade quickly. It is relatively unreactive and is stored in the fat tissue of animals. This accumulation problem is exaggerated when one organism high in the food chain eats many other ones lower in the chain. The DDT levels in the higher organisms are compounded to higher levels because they acquire all the DDT from the lower organisms they eat.

33. About 33 % to 40 % of food crop production is lost to pests each year worldwide. The monetary value of these losses is estimated at $ 20 billion per year.

35. The three structure classes of insecticides are chlorinated hydrocarbons, organophosphorus compounds, and carbamates. An example of each is shown below.

Malathion, organophosphorus compounds   DDT, chlorinated hydrocarbons   Carbaryl, carbamates

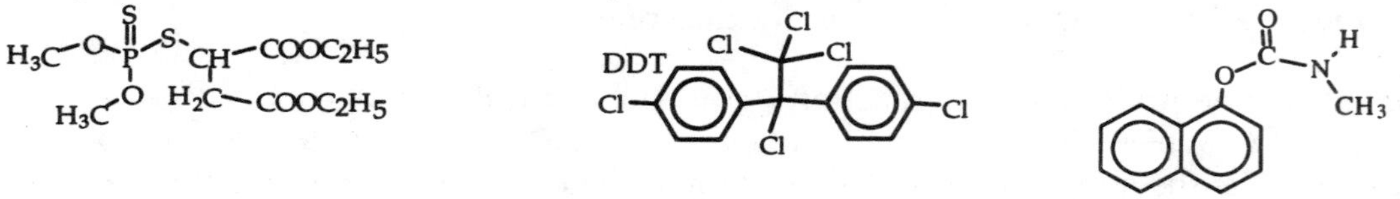

37. A biodegradable pesticide is one that is quickly converted to harmless products by microorganisms and natural processes. Pyrethrins are an example of a class of biodegradable insecticides. An example of a pyrethrin is shown here.

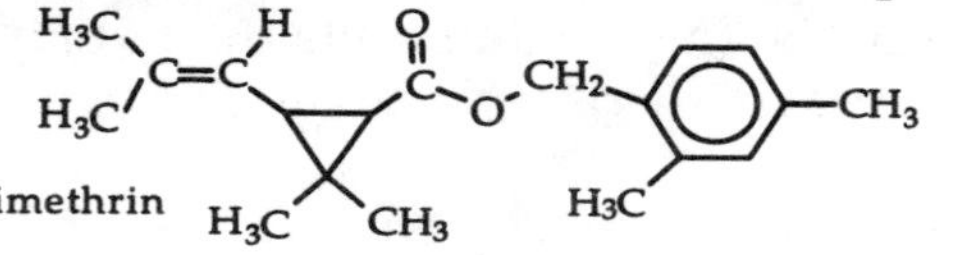

Dimethrin

39. a. Pesticides can increase crop yields and protect them when stored. They reduce the loss of crops to pests. Every bit of food saved is food that need not be replaced. Food supplies increase and food is more plentiful at lower costs. Malnutrition and starvation are reduced. Diseases carried by pests can be controlled or even eliminated. Epidemics of some diseases can be prevented or stopped. Pesticides can cause problems of water and soil contamination if they are misused. Pesticide residues can contaminate crops and create long term poisoning problems. Resistant strains of pests develop and require higher levels of pesticides.
   b. Pesticides should be used early enough to require smaller amounts and only when absolutely necessary and when alternative methods do not exist.

41. Organic farming uses materials that require only 40 % of the energy that is used in farming with synthetic chemical fertilizers. Organic farming uses human labor which has a high energy cost.

43. Answer depends on person's views.

45. *Bt* toxins are natural insecticides produced by *Bacillus thuringiensis*. Neem-tree seeds are a source of neem seed oil which is effective against approximately 200 insect species. Both are specific for pests and do not harm desirable insects. There are a number of different *Bt* toxins. Each is effective against specific insects. *Bt* toxins biodegrade in a few days. They are not hazardous to beneficial insects or other living things. Neem-tree seed oil has proven effective against pests and it spares insect-pest predators.

47. Glyphosate and sulfonylureas inhibit plant enzymes that catalyze the synthesis of essential amino acids. Plants typically synthesize all 20 amino acids. Animal species cannot produce the 10 essential amino acids in sufficient quantities for growth and must take in these as part of their diet. This means glyphosate and sulfonylureas can disrupt plant growth without posing a threat to animals and humans.
The essential amino acids are listed below.

arginine	lysine	methonine	tryptophan
histidine	isoleucine	phenylalanine	valine
leucine	threonine		

49. No right answer exists. Each person will have their own priorities.

51. The United Nations Environmental Program report filed in 1992 stated that 4.84 billion acres of cropland were lost due to overgrazing, deforestation, and soil erosion.

53. Time intervals are tremendously different. Natural breeding methods can take years while gene splicing can be done within one year. Gene insertion techniques can produce properties that are more specific and do not occur in any existing species.

Answers for Selected Problems

2. The answers are 83 % K , 17 % O. These answers come from the following analysis. The molar mass for potash , K_2O , needs to be calculated. This is done by adding the mass of K and O in K_2O.

$$
\begin{aligned}
2 \text{ moles K} &= 2 \times 39 \text{ g K} &&= 79 \text{ g K} \\
1 \text{ mole O} &= 1 \times 16 \text{ g O} &&= \underline{16 \text{ g O}} \\
1 \text{ mole } K_2O &= && 94 \text{ g } K_2O
\end{aligned}
$$

The % K in K_2O = $\dfrac{78 \text{ g K}}{94 \text{ g } K_2O}$ x 100 = 82.9 % = 83 % (2 significant digits)

The % O in K_2O = $\dfrac{16 \text{ g O}}{94 \text{ g } K_2O}$ x 100 = 17 % (2 significant digits)

3. The answers to this question are 10 % N ; 2 pounds. The label 10-20-20 means the fertilizer is 10 % N, 20 % P as P_2O_5, and 20 % K as K_2O.
The 20-pound bag is 10 % N by weight. The 10 % means there are 10 pounds of N for every 100 pounds of total weight. The pounds of N are therefore

$\dfrac{10 \text{ pounds of N}}{100 \text{ pounds total}}$ x 20 pound bag = 2 pounds of N

Speaking of Chemistry

Name _____

Feeding the World

Across

1 Name for the anion PO_4^{3-}
2 Abbreviation for integrated pest management.
5 Fertilizer containing only one nutrient.
8 Decomposed organic matter in soil.
10 Fertilizer with formula, $(NH_2)_2CO$.
15 Nutrients required in very small amounts.
16 Mixture of mineral particles, organic matter, water and air.
18 Water _____ into the structure of particulate matter.
19 Nitrogen, phosphorus, and potassium are _____ nutrients.

Down

1 Natural pesticide.
3 K_2O
4 Calcium carbonate, $CaCO_3$.
6 Genetically altered crops.
7 Chemicals used to control pests.
9 Layer of soil that is inorganic, salts, clay particles.
11 Kind of pesticide that persists and breaks down slowly.
12 First chlorinated organic insecticide.
13 Condition with pH above 7.
14 Plants that can fix nitrogen.
17 Topsoil that is friable and has a high air content.

Bridging the Gap

Name _____

Feeding the World
Internet Links To The Current State

This Bridging the Gap offers you a choice of activities on the Internet. One activity takes you into the United States Geological Survey data on agricultural chemical use and farming practices in the U.S. A second possible activity checks the Global Food Watch which is a monthly monitor of international agricultural policy.

Agricultural practices in the United States

If you choose this activity you are supposed to open the USGS page with the following URL.

http://h2o.usgs.gov/public/pubs/bat/bat000.html#HDR6

This URL gives you a very impressive page. The report consists of estimates for all counties in the conterminous United States of the annual use of 96 herbicides in 1989; annual sales of nitrogen fertilizer, in tons, for 1985-91; and land use, chemical use, and cropping practices in 1987. The information in these coverages can be used in estimating regional agricultural-chemical use across the United States. You should scroll to the following line entry and click. <u>Figure 1. Estimated annual county-level herbicide use, 1989: (a) atrazine; (b)alachlor.</u> This line has the following URL. This URL gives a map of the United States.

http://h2o.usgs.gov/public/pubs/bat/fig1.gif

The map shows the usage of atrazine in pounds per square mile for every county in the United States.

You are supposed to find your county or one nearby, read the level of atrazine herbicide use and record the amount on the Concept Report Sheet. Examine the map showing atrazine use and determine which region of the 6 below used the greatest number of pounds of atrazine per square mile. Identify the region that used the least atrazine per square mile.

Southwest   Central Midwest   Northeast   Northwest   Northern Midwest   Southeast

Additionally, your assignment is to open the following URL to access another U.S. map. This one shows the usage of potash fertilizer in pounds per square mile for every county in the U.S.

http://h2o.usgs.gov/public/pubs/bat/fig6.gif

You are supposed to find your county or one nearby, read the level of potash use and record the value on the Concept Report Sheet. Additionally determine which region of the country used the greatest number of pounds of potash per square mile. Identify the region that used the least potash per square mile.

Global Food Watch

A monthly monitor of international agricultural policy, production, prices, trade and food security. Produced by the Institute for Agriculture and Trade Policy. You use the following URL for this page.

http://www.igc.apc.org/iatp/agriculture.html

After you open this page find a topic that interests you and summarize its content.

357

Bridging the Gap
Concept Report Sheet

Name _____

1.   **Agricultural practices in the United States**
Herbicide use
What is the level of atrazine use in your county or one nearby? Record the amount here.

Which region of the country used the greatest number of pounds of atrazine per square mile.

Southwest   Central Midwest   Northeast   Northwest   Northern Midwest   Southeast

Identify the region that used the least atrazine per square mile.

Southwest   Central Midwest   Northeast   Northwest   Northern Midwest   Southeast

Which part of the country seems to have a greater problem controlling weeds? What do you think causes this?

Fertilizer use
What is the level of potash (K_2O)use in your county or one nearby? Record the amount here.

Which region of the country used the greatest number of pounds of potash per square mile.

Southwest   Central Midwest   Northeast   Northwest   Northern Midwest   Southeast

Identify the region that used the least potash per square mile.

Southwest   Central Midwest   Northeast   Northwest   Northern Midwest   Southeast

Which part of the country seems to have a greater problem maintaining potassium in soil? What do you think causes this?

2. Global Food Watch

What topic reported by the Global Food Watch interests you? Summarize its content here.

Solution to Speaking of Chemistry crossword puzzle

Feeding the World

Across:
- 1. PHOSPHATE
- 2. IPM
- 5. STRAIGHT
- 8. HUMUS
- 10. UREA
- 15. MICRONUTRIENTS
- 16. SOIL
- 18. ABSORBED
- 19. PRIMARY

Down:
- 1. PYRETHRUM
- 3. POTASSIUM
- 4. LIME
- 6. TRANSGENIC
- 7. PESTICIDE
- 9. SUBSOIL
- 11. HARDPAN
- 12. DDT
- 13. BASIC
- 14. LEGUM(E)
- 17. LABA...